FORWARD INTO BATTLE

FORWARD
INTO
BATTLE
PADDY GRIFFITH

FIGHTING TACTICS FROM
WATERLOO TO THE NEAR FUTURE

PRESIDIO

The author wishes to make it absolutely clear that responsibility for the ideas and opinions which follow lies not with any of the acknowledged persons, nor with the British Government or Ministry of Defence, but solely, entirely and completely with himself.

First published in 1981 by Anthony Bird Publications Ltd.,
Chichester, Sussex, England

This revised and updated edition first published in England in 1990 by
The Crowood Press, Gipsy Lane, Swindon, Wiltshire SN2 6DQ, England

Published in 1991 in the United States by
Presidio Press, 31 Pamaron Way, Novato, CA 94949

Library of Congress Cataloging-in-Publication Data

Griffith, Paddy.
    Forward into battle : fighting tactics from Waterloo to the near future /
Paddy Griffith. —Rev. and updated ed.
      p.   cm.
    Includes bibliographical references and index.
    ISBN 0-89141-413-4
     1. Tactics—History—19th century. 2. Tactics—History—20th
century. 3. Military history, Modern—19th century. 4. Military history,
Modern—20th century. I. Title.
    U165.G69 1991
    355.4'2'09034—dc20   90-19248
                                       CIP

Line-diagrams by Sharon Perks (pages 123, 124, 125 and 132)

Printed in the United States of America

Dedication

For Geneviève and Robert

Contents

Acknowledgements

In the original edition I expressed particular gratitude to Professor Norman Gibbs, and also to Ned Wilmott, Andy Callan, John Ellis, Christopher Duffy, Nigel de Lee, Richard Holmes, David Chandler, John Keegan, Ann Johnson and John Hunt – and I wish to repeat and reinforce all that gratitude here.

In the years since 1981 I have been quite overwhelmed by the scale of the additional help and support which I have received from many other quarters. I cannot, alas, mention everyone here, and I am sincerely sorry that this listing is necessarily so uninformative and incomplete. I owe especial thanks to Elmar Dinter, Andy Grainger and Paul Harris for many stimulating discussions on these subjects; but scarcely less important has been the help freely given by the following additional persons (in alphabetical order, and with apologies for omission of titles, degrees, military ranks, and so on): Martin Alexander, Dick Applegate, Jim Arnold, Diana Ashdown, de Witt Bailey, Ian Beckett, Matthew Bennett, Kathy Clarke, Tony Clayton, Peter Cross, Mark Cuthbert-Brown, Peter Dennis, Charles Dick, Chris Donnelly, John Drewienkiewicz, Charles Esdaile, Hugh Faringdon, R.E.M. Foster, Steve Fratt, Michael Glover, Ian Greenwood, Arthur Harman, Paul Heap, Peter Hofschroer, Sidney Jary, Barrie Jones, Chris Kemp, David Kirkpatrick, John Koontz, Jean Lochet, Kenneth Macksey, Peter MacManus, Terry Martin, Greg McCauley, Phil Melling, René and Marguerite Mouriaux, Anne Nason, Andrew Orgill, Joe Park, Dan Radakovitch, Gunther Rothenberg, Richard Shirreff, David B. Sweet, Wally Simon, David Swinburne, Freddie Walker, Jeff Walsh, Howard Whitehouse, Patrick Yapp and Ned Zuparko.

I am particularly grateful to the librarians at both Sandhurst and Camberley for their diligence; to the members of and speakers at the 'Wardig' society for their searching weekly analyses; to all my RCC and JCSC students since time began, for their unconscious yet very real inspiration; and to Tony Bird for his all-too-often-tested forbearance.

Paddy Griffith
Owlsmoor, August 1989

ix

Preface

This second edition of *Forward into Battle* has been significantly expanded and revised since the book first appeared in 1981. The main chapters have been retained almost unchanged, but new sections have been added at the end of each one, and both the introduction and final chapter have been rearranged and expanded. The extra material is intended to update the original findings at the same time as exploring the subject matter in a little more breadth and depth.

Since 1981 I have looked at several additional aspects of the history of modern tactics, and have filled in some of the gaps in my story that were then identified by readers. There are therefore now additional discussions of Napoleonic firepower, of the American Civil War, of the French in August 1914, and of the use of armour both during and after the Second World War. Furthermore, the story no longer stops at 1973, but has been extended into 'the near future'.

When it was originally published in 1981, *Forward Into Battle* tried to break new ground in a number of different directions. It challenged, for example, the generally accepted view that Wellington's infantry had won its victories by firepower from a static line. It revealed that 'the empty battlefield' was not quite the early twentieth-century innovation that had often been assumed, but had earlier precedents, and it questioned the popularly held stereotype that the tank dominated the battlefield between 1916 and 1945. It also opened up an unconventional perspective on the 'helicopter revolution' that has often been claimed for the American mainforce effort in Vietnam.

The book was greeted with enthusiasm and great interest by some, but with a certain incredulity by many others. The idea that technology and 'hardware' were not the be-all and end-all of combat seemed at that time to be such a radical and strange perspective that many people were tempted to dismiss the book out of hand. Veterans of Vietnam who had seen other aspects of that war came close to denying the reality of the firefights in the Central Highlands reported by S. L. A. Marshall. 'Tankies' defended the reality of the allegedly all-conquering, all-tank *Blitzkrieg*. First World War buffs blustered their denial of Rommel's

1914 experience of innocuous shelling, and Wellingtonians wilfully claimed that 'Griffith believes bullets do not kill people in battle'.

It is, of course, gratifying for any author to attract such a sparkling spectrum of criticism, since 'all publicity is good publicity'. However, it is also frustrating to find that one's long-considered and documented findings, on such a wide spread of subjects, can apparently be dismissed so lightly and easily. Almost immediately following their publication, however, the general ideas expressed in *Forward Into Battle* chanced to be tested against the reality of a new infantry war – this time one fought for possession of the Falkland Islands in the South Atlantic Ocean.

I certainly experienced a measure of the tragedy and irony of the 1982 events, since a number of my former students and friends were direct participants, and others were significant 'helpers'. I was moved by the vivid picture of modern warfare which the survivors painted – but I also found grounds for believing that my own earlier 'unfashionable' propositions had been quite vividly, pungently and appropriately vindicated. Whatever else it was, the Falklands War was surely a triumph of morale, training – and cold steel – over numbers, 'hardware' and bombs.

Still more telling, perhaps, were the subsequent radical shifts of emphasis in both the American and British operational doctrines for a possible future war on NATO's central front in Germany. The Americans seemed at long last to have interpreted the lessons of the Vietnam mainforce war in the sense of 'morale and manoeuvre', and to have glimpsed some of the limitations of their time-honoured 'firepower' and 'managerial' approach. The British, for their part, were finally starting to shake free from some of the more debilitating rigidities of the Haig and Montgomery traditions.

It cannot be claimed, alas, that my little book was in any sense responsible for any of these historic changes in military policy – but I think it can at least be seen as having been abreast of the times before 'the times' had fully occurred.

1

Introduction

Nicky Rokossovsky held his breath as his gunner's laser rangefinder registered on the blurred image of the T94, then the protracted muffled 'boom' of firing told him that the first 120 mm round had started on its three kilometres of accelerated, arching flight. The unequal midnight battle, for which he had manoeuvred his squadron so hectically during the past four hours of darkness and radio silence, had finally begun. Everything now depended on whether his tiny force of eight Challenger Mark VIIIs could make its ambush stick against that Soviet tank battalion eerily jinking its way across the video-image of meadowland in the distance – and then whether he could bug out in time to find and maul the second battalion, which he knew was trying to come in from behind.

Ironically enough, Nicky was himself a distant relative of a hero of the Soviet Union in the Great Patriotic War – a Marshal who had commanded whole groups of armies, no less. His parents had led a protected, cosmopolitan life in Moscow until forced to flee West during the post-Gorbachov convulsions and purges. Then within a year their son Nikolai had been born and now – after Eton, Sandhurst and Bovington – here he was as a junior captain in one of the smartest cavalry regiments of the British Rhine Army, facing the Warsaw Pact's surprise attack on Christmas night, 2020 AD. He had originally been mightily vexed to be made duty officer for the Christmas holiday – but now he was far too busy to worry about such things. Nor did it ever occur to him that he, Nicky Rokossovsky, was fate's selected representative of four long generations of vengeful East European émigrés. He, like the White Russians of the 1920s or the Poles and Hungarians of the 1940s and 50s, had volunteered for British service so that he could fight back against a Communist system that had run off the rails – but unlike them, he was now being granted a chance actually to do the job.

With a fleeting sigh of relief he watches his squadron's fire take effect, and the enemy tanks start to brew. Now is the time for rapid action. Quickly turn on the lasers to blind the enemy image intensifiers; engage the follow-on targets with a second and third salvo of kinetic energy

1

rounds; flash out the squadron code to break contact; release a laser-defeating smoke cloud, then rev up the mighty turbines to disappear backwards into the trees behind the ridge. Scour the ever-shifting wavebands to report back directly to the brigade command post – while regimental HQ is apparently still disentangling itself from barracks (and probably suffering from a collective festive hangover). Then wait anxiously for the vital drone surveillance screen to come to life, and definitely locate that ominous second enemy battalion.

Damn! You can't hear yourself think, because the whole of the Red Army's artillery seems to be demolishing the crest from which we have just retreated. Can't they stop that noise for just an instant? At least it shows we're still one jump ahead of their famous target acquisition system – but we must get moving quickly now. Damn and damn again – Tango Three has just been caught in the cone of incoming fire, and has collected a couple of smartlets through the engine mantle. . . . Can't stop here to pick them up – Jimmy and his crew will have to walk home.

OK, the enemy vectoring is now flashing up fuzzily on the video screen: let's think about finding an ambush position and checking out our approach to it. Take the lead in the squadron column of march, and concentrate hard on the map reading. Re-impose radio silence and activate the stealth technology. Send off Corporal Ames to do his stuff in the second ECM decoy Scimitar. He must be the brainiest man in the squadron – but will we ever see him again? Too late to fuss about that now – the key thing is to land our blow before the enemy gets a sniff of us. Driver advance! . . . and gunner keep your hand on the chaff dispenser!

This passage gives us a glimpse into a major land battle as it may be fought at some point during the next generation. It suggests that action may start with a complete surprise, then take place at breakneck speed, 24 hours a day, using plenty of electronic gadgetry and some devastatingly high-powered vehicles and weapons. There will be novel features in abundance – yet below the surface there will also be much that is conceptually familiar. Small groups of tanks will fight fleeting duels of manoeuvre and rapier-thrust, in the knowledge that to halt for long in any one position will inevitably attract a thunderstorm of indirect fire. Command and communication will be intermittent, to say the least, and much will depend on individual qualities of leadership, training, and quick thinking at the lowest levels. Battles will be fought in a fog of tension, uncertainty and confusion – just as they always have been, in fact.

In any future war we may well expect to see 'encounter battles' fought between relatively small forces. Columns of tanks and APCs will bump into each other off the line of march, and fight combats that have not been planned. Both sides will expect to excel in this type of engagement, in which the clever manoeuvre of armoured forces will be decisive, but neither side will really know much about how the enemy reacts until it is too late. No one will be able to perfect their small unit 'encounter' skills until they have already experienced a number of such engagements.

Armies do not change their overall shape very quickly, and least of all in times of *détente* and economic stringency. They tend to display a continuity in tactical thinking which is reinforced by long lead times in weapon production and financial restrictions on procurement. For example, as late as 1945, two world wars after 'the taxis of the Marne', much of the German army's transport was still horse drawn. As late as 1990, just half a century after flak had decisively secured Guderian's Meuse crossings, the British army still lacks a self-propelled, rapid-fire AA cannon. By the year 2020, therefore, it is not unreasonable for us to suppose that a continuous line of development may still be strongly evident from the weapons and tactics of the past, or that the many predictions of a completely electronic, remote-controlled, space age battlefield will still remain largely unfulfilled.

With this in mind, it is the purpose of the present volume to analyse the continuous development of the minor tactics and weapons of land warfare from Napoleonic times to the near future, not forgetting the often misleading interpretations that have been placed upon it by analysts and historians. The book will discuss a variety of historical engagements and combat doctrines, and will go on to make speculations about the state of tactics today and tomorrow. It does not concern itself with the political or strategic reasons for which battles were fought, nor with the decision-making of war ministers and field marshals. Instead, it looks at what is supposed to have happened, and what really did happen when two opposing groups of soldiers came face to face across 'no-man's-land': the lethal but crucial area where the outcome of combat is decided.

No-man's-land may be defined as the zone between the forward edges of two armies which are locked in combat against each other. It is usually an uninhabited and desolate place in itself; but on either side of it there will be bodies of armed and frightened men who are fighting for their lives. These bodies will frequently exchange shots at long range, and may from time to time attempt a forward movement to finish off their opponents at close quarters. The outcome of the greater conflict will usually revolve around the success or failure of a multitude of such attacks. If the defenders can by and large keep their perimeters inviolate they will triumph. If they fail, then the attackers will be free to move on to fresh victories further afield. It is thus the struggle for no-man's-land which determines the shape of the battle as a whole.

The physical extent of no-man's-land has varied greatly from one period of history to another. In ancient times it was determined by the reach of the front-line soldier's personal weapon; perhaps no more than a few feet in the case of a sword or spear, or a couple of hundred metres for a bow. In the course of the last two or three centuries, however, there has been an astonishing increase in the range and power of weaponry which has brought more complex factors into play.

In eighteenth century and Napoleonic warfare armies were often kept apart by the field artillery, which could reach out to a kilometre or more. Around the third quarter of the nineteenth century this range was extended by several additional kilometres with improvements in both guns and ammunition. Since that time direct-fire weapons have been constructed which can destroy any target at a range of four or five kilometres; while indirect-fire or aerial weapons can now take effect at tens, hundreds or even thousands of kilometres further afield. The extent of no-man's-land has thus been called into question during the past few generations.

In principle we would expect to find that no-man's-land has grown in width in proportion as weapon ranges have increased; and it is true that in many modern battles the two armies have seemed to glower at each other across a very wide strip of country indeed. In desert conditions, or on the broad plains of Eastern Europe, there are few terrain features to give protection at close range. It is only by putting distance between oneself and the enemy that any degree of security can be achieved.

In other cases, however, the practical effectiveness and range of weapons has been limited by terrain features. Soldiers have been unable to see as far as they could shoot, and so they have been incapable of hitting an enemy lurking at no great distance from their own positions. In towns such as Caen or Stalingrad, or in the jungles of Burma and the Pacific Islands, fierce battles were fought during the Second World War with only a few yards separating the combatants. There has sometimes also been a paradoxical tendency for troops to seek safety from long range weapons by 'hugging' the enemy positions as closely as possible. Some trenches in the First World War were dug to within fifty metres of each other, for example; while in Vietnam engagement ranges were often no more than thirty metres. It is therefore quite misleading to suggest that combat ranges have grown inexorably greater as weapon power has increased.

The breadth and depth of the battlefield, on the other hand, have both certainly increased a very great deal. In Napoleonic times an army could occupy a space no more than a few kilometres wide by two or three kilometres deep, and troops knew they would be safe from enemy fire as soon as they moved behind the nearest hill. In the twentieth century, by contrast, armies have spread sideways until they fill literally any breadth of front available. Their depth has increased to tens of kilometres in

4

many cases, and may be made up of successive belts of mines, clusters of strongpoints, and finally a line of reserve positions. Modern armies also require extensive supporting organisations behind them, and these in turn can be brought under long range enemy fire or air attack. It has become a feature of modern war that an interdiction battle will be fought behind the front line with almost as much ferocity as the front line battle itself. This is a form of tip-and-run or standoff fighting, in which raids and bombardments take place without the permanent occupation of ground.

In this book we shall not be looking closely at these interdiction battles behind the lines, important though they have undoubtedly become. Instead, we will be concentrating our attention on what goes on further forward, where troops move under direct fire to take ground from the enemy. This necessity is what gives no-man's-land its special quality. It is the arena of personal confrontation between the two sides, where the abstract logic of higher commanders and staff officers suddenly takes on a very much more concrete form.

In order to identify what actually happened across no-man's-land during the past two centuries, we must look at the evidence which has come down to us. This is largely written evidence in one form or another[1]; and it often turns out to be in disappointingly short supply. Minor tactics seem to be less well covered in the documents than are many other military activities. This is hardly surprising, perhaps, since tactics do not routinely generate paperwork in the same way as do such tasks as intelligence gathering, logistic stock-keeping or staff planning. It does not come naturally, after all, to record the minutes of a firefight when you are sprawled across a muddy field being shot at.

Most of the written sources which discuss tactics do so at second hand. There are more commentators in this subject than there are direct observers. Nor do either of these classes of writer tend to be particularly helpful. Whereas the commentators suffer from an inordinate measure of political or romantic bias, the survivors of battle tend to give us only the undigested and fragmentary scraps of information which they have been able to collect from their own personal foxholes. There is often but scant agreement between different authorities, and precious little solid scholarship. Only rarely do we find a piece of tactical analysis which is convincing and clear.

It is for these reasons that the history of tactics has too often been misrepresented. In the pages which follow we shall therefore attempt to correct the record over certain important issues which seem to have been garbled rather than clarified by the benefit of hindsight. We shall be taking up the cudgels against those who have given us a false picture of what really happened. In particular we shall concern ourselves with three specific groups of authors, each of which transmits a different type of evidence about tactics. These three groups may roughly be classified as 'journalists', 'analysts' and 'survivors'.

In the category of 'journalists' we include war correspondents, official propagandists, and subsequent historians who are writing for a non-specialist general readership. These people will not normally be concerned with the details of minor tactics, and will usually allow them to be overshadowed by higher and weightier matters. They will know, for example, that the general reader is more interested in Napoleon's strategic decisions at Waterloo – or even in what he ate for breakfast – than in the precise method by which his Old Guard was routed. If they have a choice between truth and legend, furthermore, writers of this type will often prefer to print the legend: it makes a better story and sells more copies.

It perhaps goes without saying, and even without reading Clausewitz, that wars are usually reported in a heavily political manner. Extreme political passions are always aroused by hostilities, and every casualty increases the emotional level still further. In these circumstances objectivity tends to fly out of the window. In 1915, for example, Rudyard Kipling wrote in the *Morning Post* that 'There are only two divisions in the world today: human beings and Germans.'[2] This sort of attitude scarcely makes for careful reporting of minor tactics, and we find that wishful thinking frequently creeps into battle descriptions.

The most obvious example of the process comes in the reporting of relative losses. It is almost a standard practice to estimate that the enemy has suffered at least double his true casualties, while at the same time halving your own. This tendency was already very prominent in Napoleon's battlefield bulletins. It appeared again in the official histories of the First World War, while in Vietnam the MACV 'body-count' became notorious for its margins of error. It is perfectly true that losses are very difficult to gauge exactly in the heat of battle: but that does not often seem to have prevented confident official statements from being made. Once they have been reported in print they tend to be repeated rather uncritically by subsequent generations of writers. They are very hard to correct later on.

What is true of combat losses is also true of tactical methods themselves. Once the public has got it into their heads that Wellington's infantry fought primarily by musketry fire, for example, or that Asian infantry fought primarily by 'human wave attacks', then it becomes extremely difficult to change the history books. It is much easier for historians to repeat the popular image of how these battles were fought, and pass on quickly to the apparently 'more important' or 'more interesting' questions of strategy and military biography.

When we turn from the 'journalists' to the 'analysts' of minor tactics, we find a rather more serious type of commentator.[3] These men are looking at past battles in order to draw operational lessons for the future. We will expect them to be more objective and less influenced by political propaganda or public tastes. They have an obligation to use accurate facts so that they can devise truly effective fighting methods for their nation's soldiers in the next war. It is their job to create sound tactical theories upon which training manuals can be written and tactical leaders indoctrinated.

It is true that the standard of such commentaries is usually higher than that of the writers for the general public; but analysts also suffer from certain occupational diseases of their own. In the case of purely academic tactical historians, there has been a distinct tendency to concentrate excessively upon the items of 'hardware' that can be quantified. Such things as weapon specifications, theoretical drill formations or marching speeds have been scrutinised in great detail, as potential keys to what happened in battle. In the process, however, historians have often missed the ways in which actual battlefield practice differed from theoretical performance. Imponderables such as the level of training, fear and morale, for example, often exercised a quite disproportionate effect on the ultimate outcome – and the failure to understand these factors is frequently a major source of distortion in the historical record.[4]

Most official analysts, on the other hand, do not enjoy the relative independence of a detached academic environment, but work as members of an army. They are subject to the many institutional pressures peculiar to armies, and in particular they will find themselves within a rigid hierarchy of ranks, where it is often rather difficult to question the decisions of superior officers. Their findings will also tend to be discussed and re-worked in endless committees, with the injection of many political considerations extraneous to the subject-matter itself. In particular there will be pressure groups from different arms of the service to put forward conflicting departmental views, and there will be strong budgetary influences not far from the surface. All this may finally conspire to distort the objectivity of analytical findings.

Institutional pressures within armies often lead to the formation of tactical 'schools', each of which is fiercely committed to one particular battlefield technique. In the late eighteenth century French Army, for example, there was a bitter and rather sterile debate about whether infantry should attack in columns for shock action, or in lines to develop their firepower.[5] Each side in the debate wanted the kudos which would come from official sponsorship of their particular tactics, and so the arguments tended to become more and more extreme. They often lost touch with the hard realities of the battlefield, and drifted off into wonderous theoretical constructions which would have been quite impractical in war. In the end the quarrel was deemed to have been resolved by Napoleon himself, who produced an equally unrealistic

7

compromise solution known as the '*ordre mixte*'. He often recommended it to his generals; but they regarded it with suspicion and did not often use it in battle. They felt, perhaps, that as a former artillery officer Napoleon was making an altogether too bookish and academic approach to infantry tactics.

This particular debate makes an interesting illustration of how tactical analysts can become side-tracked into unrealistic and doctrinaire systems. It is by no means unique. Another example comes in the nineteenth century, where the tactical debate became polarised between the believers in the improved firepower of the age, and those who felt that the bayonet attack was still both practicable and profitable. The former invoked the supposed powers of an unproved technology, while the latter pointed to the real outcome of a thousand actual battles. The believers in the bayonet have nevertheless been unjustly vilified as dreamers and 'military spiritualists'; whereas the champions of firepower have recently been accepted as profound and original thinkers who had the general welfare of mankind as their prime concern. In fact, of course, both sides of the debate found themselves being pushed to extremes by the pressure of the opposition. Institutional factors once again blew the analysts off course.

In the present century we can find plenty of other 'schools' of tactical doctrine. In the 1920s and 1930s, for example, the prophets of mechanised mobility found themselves resorting to shrill extremism in their vain attempts to gain official acceptance. Their basic case was a good one; but they overdid its presentation and once again came to abuse their historical raw material. Another example might be found in the American forces of the 1960s, where a veritable parliament of opposed tactical schools sprang up. There was a Marine Corps school, which made no less than sixty amphibious beach assaults in Vietnam, in the hope that the enemy would oblige by opposing them. There was an army helicopter lobby, which saw the airmobile division as the answer to all ills. There was an electronic battlefield establishment and an armoured warfare establishment. There were riverine forces, special forces and, of course, air forces. Each of these groups analysed the tactical problem in different terms from the others. They drew upon different data and extrapolated different trends into the future.

The tactical analyst is necessarily working at a considerable distance from the battlefield he hopes to influence, and he is usually forced to simplify his case in order to 'sell' it. This tends to focus the debate around somewhat extreme or over-theoretical propositions or phrases, in which one individual abstract quality such as 'morale', 'mobility' or 'firepower' generally comes to be seen as predominant over the others. Tactical history can then be re-written by the proponents of that particular quality in order to support their case. Not until a system has been discredited on the battlefield will some new wave of analysts arrive to demonstrate the vital importance of a different abstract quality altogether.

If historians follow the expectations of their public, and analysts follow the abstract logic of their 'schools', what can we make of the more direct evidence of battle survivors? These are people who have actually lived through the fighting they describe, and they will bring vivid memories to their writing. It is true that in some cases these memories may be garishly embellished, and in others they may be unduly reticent; but by and large we ought to be able to accept them at more or less face value.

The difficulty with the reports of survivors is rather that they are fragmentary. No one man in a battle can see very much of what is going on, nor will he have leisure to weigh the significance of what he does see. As Sir Ian Hamilton put it: 'The worst of writing on a battlefield is the necessity which it entails of constant contradictions.'[6] If a coherent and balanced narrative of the action is to be written, a considerable effort of comparison and analysis will be required between the reports of many different witnesses. This task is undertaken only too rarely, and it is followed through to completion more rarely still. Thus Captain Siborne collected a splendid wealth of eyewitness accounts of the Battle of Waterloo; but he failed to resolve all the contradictions between them, and arrived at a final picture which appears to contain a number of important inaccuracies.[7] It is only with the work of General S. L. A. Marshall during and since the Second World War that the technique of combat history has been put on a sound basis. His descriptions of the combats in Vietnam, for example, are models of the genre.[8]

The vast majority of battles leave us but few useful eyewitness accounts, and it is rare for any two observers to describe exactly the same event. Nor do they always tell us the sort of details which we would most dearly like to hear about. A participant in a battle may find some things so obvious that he will not bother to record them; and yet these are often precisely the aspects which remain obscure to the distant historian. For all that, however, it is from eyewitness accounts that we can gain our best and most direct impression of what really happened. Provided that we approach them with due caution, we can glean a lot more details from this type of writing than we can from the work of the 'journalists' or the 'analysts'.

Throughout what follows we will be constantly on the lookout for the traps and pitfalls into which the written evidence may lead us. We will try to see what political or romantic slants may be concealed behind 'generally accepted' views; and we will try not to accept the generalisations of historians until they have been supported by some solid eyewitness reports. We will also watch for the institutional pressures at work upon analysts in the tactical debate which may corrupt or distort their vision. Finally, we will try to piece together and make a coherent story out of a few of the fragmentary memories of battle which have come down to us from those who were actually present.

All this amounts to rather an ambitious task; but it will be worthwhile if it can help us towards a better understanding of what actually went on in combat. There is at present too much which remains obscure in this subject, and any new light which can be brought to bear will be long overdue.

A particularly thick pall of obscurity seems to hang around the role of technology and 'firepower' in battle; and these two factors have attracted a disproportionately high degree of misunderstanding from commentators. It has been claimed, for example, that as early as 1808 Wellington's infantry was habitually winning battles purely by its musketry fire. We are invited to believe that 'solid' British infantrymen could already, one hundred years before Mons, lay down such an impenetrable hail of shot along their frontage that enemy attacks would be simply blown to pieces. This curiously anachronistic image is doubtless a source of great comfort and pride to many patriotic Britons, and it has found its way into most of the history books. As we shall see, however, it is totally misleading.

We are also often told that the rigid and brainless military hierarchies of the nineteenth-century failed to realise what important technical changes were being made under their very noses. When a devastating new generation of weapons was unleashed in 1914 it is supposed to have come as a traumatic surprise to everyone concerned. Awareness of the so-called 'empty-battlefield', in which soldiers are forced to disperse and dig for shelter, is considered to have been an early twentieth-century, or at best a late nineteenth-century phenomenon. In these pages, however, we shall attempt to demonstrate that it was really an early nineteenth-, or just possibly a late eighteenth-century event. The debate about the empty battlefield was of very long standing indeed, and the doctrinal answer to it had already been formulated, in quite specific ways, at least a hundred years earlier.

Although 'the empty battlefield' had long existed, it was never to be much more than one among several different forms of combat. In the American Civil War it was assumed to be predominant from the start – and that expectation has deeply coloured the subsequent historiography – yet it was seen in practice far less frequently than has been assumed, at least until the year of exhaustion, 1864. Far from being unaware of improved firepower in 1914, therefore, the general staffs were heartily sick of being told about it. They had heard the cry of 'Wolf!' only too often before.

Continuing our story into the twentieth century, we find that improved firepower did eventually outrun the tactical counter-measures that were thought to be available. While the true solution should have been the creation of a 'new infantry' and a more flexible artillery, the armies remained fixed in a deadlock which could supposedly be broken only by new technology, in the form of the tank.

A similar tale can also be told of the American mainforce war in Vietnam. The tactical deadlock continued; and not even the vast resources of modern technology seemed capable of breaking it. Sophisticated systems for surveillance, mobility and firepower alike could in practice deliver no more than a fraction of the benefits which they had originally promised. The traditional role of the infantryman, despite everything, remained paramount.

A great deal of this has been ignored or obscured by the many commentators who still view twentieth-century warfare through technologically tinted spectacles. They do not look far enough into the problems of no-man's-land because, in a full sense, they have been 'blinded by science'. What we shall be attempting in the present book, therefore, is to take a more careful look at the evidence, and so rectify our picture of what actually took place in battle.

2

1808–15: The Alleged Fire-power of Wellington's Infantry

Our story starts in 1808 with the opening of the Peninsular War. This date marked the beginning of the end for Napoleon's armies, since they were thereafter forced to fight on two fronts with ever diminishing success. It also marked the final emergence of the small but professional British army as a force to be reckoned with in the continental warfare of the day. Starting with the Battle of Vimeiro (1808), which cleared the French from Portugal, and culminating in the epochal Battle of Waterloo (1815) the British seemed to go from one victory to another. Much of this success was due to the genius of the Duke of Wellington; but a great part of it was also due to the tactical efficiency of his infantry.

Wellington's infantry formed a greater proportion of the army than was normal at the time, and it bore the brunt of all the battles. It had a reputation for holding defensive positions against whatever masses the French could throw against it. Deployed in 'thin red lines'two deep (the normal practice was to form in three ranks), it would wait until the enemy attack columns came to close range, and then defeat them decisively. Indeed, it was better fed and better trained than the French in the Peninsula; but it was also heavily outnumbered and outgunned. Its successes at the point of contact are thus remarkable and indicate that this was a highly unusual force. A study of how these successes were achieved may reveal much about the tactical conditions of its age.

At first the phenomenon of the British infantry did not excite very much comment, and its tactics remained somewhat mysterious to contemporary observers. As late as Waterloo there was still only a minority of French officers who had learned to respect it at its true value, while not even the extensive post mortem in France after the wars produced very much discussion of the subject. The amazingly prolific strategist Jomini, for example, wrote no more than a few pages about it.[1]

12

What little Jomini did write is interesting both because it was influential among later commentators, and also because it was one of the first attempts to identify a tactical 'system' behind Wellington's battles. Jomini was full of praise for the 'solidity' and 'murderous fire' of the regular British infantry, and like most French commentators of his day he pointed to the advantages of British volunteer recruiting methods over Napoleon's mass conscription. He then went on to describe how at Waterloo Picton's line had enveloped the heads of the French columns with fire, to which the latter had been able to reply with only very few muskets. This implied strongly that it was the relative numbers of muskets bearing on the enemy which determined the outcome.

According to Jomini's implied interpretation, any column formation ought to have been at a disadvantage against a deployed line, since a column would inevitably have the smaller frontage of muskets. What he actually said, however, was that the French might well have succeeded against the British if only they had used small and flexible attack columns rather than large and cumbersome ones. Jomini therefore hedged his bets to some extent, and suggested that firepower was not necessarily the totally decisive factor in these battles after all. His final position turns out to be somewhat paradoxical. He suggests that solid regular infantry deployed in line can hold its ground by musketry fire alone, yet that well organised regular infantry can advance through such fire to achieve the victory. This paradox was to bedevil French thinking right up to 1914, and it ultimately had disastrous results. In terms of the debate of the 1830s, however, it was no more than an interesting aside. Little attention seems to have been paid to it. What really did happen in Wellington's battles continued to be somewhat mysterious until the very end of the nineteenth century, when the nations of Europe became fired by both a new brand of extreme nationalism and the need to make careful technical preparations for the next war. As part of the general arms race around 1900 there was a renewed interest in the last great European conflict, namely the Napoleonic Wars. Every nation conducted painstaking official and private studies which are still, today, by far the best analyses we have of the Napoleonic period.

One result of this historiographical effort was that the question of Wellington's infantry tactics was at last brought into the open. In Britain Sir Charles Oman's monumental *History of the Peninsular War* laid bare the tactical system of the two sides with the same lucid confidence which he brought to all other aspects of the struggle. In France the General Staff historical section threw up an example of that rarest of species, the true tactical historian, in the shape of Commandant Jean Colin. As a small part of his general work Colin explained what had happened in the Peninsular battles, and set it alongside the development of Napoleonic tactics as a whole.[7]

13

The matter would have rested there, had not these two great authorities chanced to disagree. In retrospect it was perhaps inevitable that they should have done so, since they were of different nationality and writing at a moment when national differences seemed to be of paramount importance. Their disagreement has nevertheless remained unresolved, and today we have two contradictory accounts of what happened when Wellington's infantry faced their French counterparts. There has never been a serious effort to find what really took place.

Oman's view is that the British owed their success primarily to their superiority of fire and protection. This derived firstly from the choice of terrain which sheltered their main infantry line from the enemy artillery, and secondly from their heavy and efficient skirmish screen which kept French skirmishers at bay at the same time as it inflicted casualties upon approaching attackers. There can be no question about either of these factors, and Oman was quite correct to underline their importance. His third point, however, was the one to which he attached the greatest importance. He said that the British superiority came from their close range fire by extended lines of infantry against dense French columns.

It is the mathematical relationship between the column and the line which is the hallmark of Oman's view. He makes, at some length, the point that a battalion of soldiers drawn up in a line two deep will be able to point more muskets at the enemy than an equal number of soldiers massed in column on a frontage restricted to between thirty and a hundred men across. One has only to count the muskets, and one will find that the firepower of the line will always be greatly superior to that of the column. It therefore follows, according to Oman, that Wellington's troops could shoot the French to pieces faster than vice versa, even if the French column was two or even three battalions in depth against a single British battalion. He also admits that the French must have dimly realised this in the course of some of their attacks, since they did occasionally attempt to deploy into a wider formation once they had become embroiled in a fire fight, in order to bring more muskets to bear.

Oman's view, supported as it was by his prodigious documentation, brought the debate onto a much more scientific plane than it had previously reached. His arguments were neat, attractive, and apparently unassailable. Unfortunately, however, they seemed to suggest that the French tacticians of the Napoleonic Wars were rather stupid, and blind to the most fundamental principles of infantry fighting. That was an idea which Commandant Colin refused to accept. In Colin's view the French had not really been trying to make attacks in column at all. They had known as well as the British that the important thing was to deploy into line to achieve maximum firepower; but their problem had been a failure to know exactly where the British were waiting for them, and thus where they ought to have deployed their own line.

Colin leapt upon Oman's admission that the French had sometimes attempted a deployment at the last minute, and claimed that this demonstrated a desire to attack in line rather than in column. Column was the formation for manoeuvre, he said, but not for combat itself. He supported this claim with a number of examples of French attacks in line drawn from other theatres of war, and pointed to the French 1791 infantry drill book, which clearly advocated line attacks with volley fire in the eighteenth-century manner. Finally, he asserted that Wellington's use of dead ground to shelter his men from artillery fire made it difficult for attacking French infantry to locate them. Attacks tended to blunder into the British line before the French had a chance to deploy into their preferred fighting formation.

The disagreement between Oman and Colin was centred upon the question of what the French infantry had intended to do at the point of contact. Oman claimed that it had wanted to charge forward with the bayonet in a massed column, relying wholly upon shock action and élan. Colin, on the other hand, denied this and said that the French had wanted to develop a musket line to beat down their opponents by firepower. It is important to note, however, that they were unanimous that the British did rely upon firepower, and that whatever the French may actually have wanted to do, they too should have relied upon firepower. Within the disagreement, therefore, there was a high degree of accord about the importance of firepower.

It is tempting for us today to accept the things about which Oman and Colin agree, and take sides only on their points of difference. Almost all modern authorities have in fact done precisely that.[3] There is something highly persuasive to the English mind in Oman's account of the thin red-coated line standing its ground against heavy odds, doggedly firing until the foreign hordes melt away. Apart from anything else, it seems to fit perfectly into the national tradition set at Agincourt and later continued at Mons. The Englishman likes to think of his soldiers as expert shots who can get far more from their personal weapons than their excitable alien opponents. Whether the weapon be a longbow, a Brown Bess musket or a Lee-Enfield rifle seems to make little difference. Phlegmatic solidity and accurate shooting appear to be factors which have remained constant through the centuries.

Another assumption which is almost universally accepted today is the general point that it must be firepower which kills people in battles. The enormous improvement in weapon power during the twentieth century seems to make this conclusion unavoidable. The butchery by direct fire weapons during the First World War has left an ineradicable impression upon all of us. When we contemplate a battle before the twentieth century, therefore, it is only natural that we should want to interpret the result primarily in terms of firepower. If writers from before the First

World War take the same line, as Oman and Colin did, we are only too ready to accept their findings at face value.

Oman's arguments are certainly alluring, but we must ask ourselves whether they may not also be a trifle anachronistic. We tend to forget that the rise of weapon power had already been noticed and had influenced theorists even in Jomini's day, some eighty years before the First World War. It was certainly a commonplace by the time Oman was writing. His geometrical count of muskets between the column and the line, in fact, looks suspiciously similar to another tactical debate which was going on around 1900, in the field of naval warfare. In ironclad naval tactics the important thing was known as 'crossing the enemy's "T"', or placing one's fleet in an extended line across the head of the enemy's column. By doing that all one's guns could fire in broadsides, while only a few of his forward-facing guns could fire from the head of his column. Precisely the same mathematics would then apply as Oman had claimed for the Peninsular battles. So he did not have very far to look for his model: it was ready made for him by the naval gentlemen of his day.

We are left with the suspicion that the 'firepower' interpretation of Wellington's success may have been accepted rather too rapidly, due to the technical developments which took place after the Napoleonic Wars had ended, as well as the existing national myth of British infantry fire accuracy through the ages. It also seems odd that so very few accounts have come down to us from the wars themselves which describe Wellington's battles in anything like the neat mathematical terms employed by Oman. Napier comes nearest, but with Jomini he is very much in the minority. In general the phraseology which eyewitnesses chose to use in their accounts seems to be rather seriously at variance with the mental constructions of later writers.

As an example of this discrepancy it is worth looking at the defeat of Thomières' 'Brigade' (actually only two battalions) by the 1st Battalion of the British 50th Regiment during the Battle of Vimeiro in 1808. Depending on which authorities one believes, this was either the very first 'column against line' action of the Peninsular War, or the second by only a short head. It is certainly one of the actions which has made a forceful impression upon historians, as is illustrated by the following passage which appeared in 1962:

The first volley from the First Battalion of the Fiftieth was fired at a range slightly over 100 yards; others followed regularly at 15 seconds intervals as the range gradually shortened. Slowly, the flanks of the 50th wrapped around the column. The British line was using every one of its 900 muskets; the French could reply with no more than 200 of their 1,200 firearms. General Thomières, who commanded the French brigade, endeavoured to deploy from column to line under fire, but found this impossible. The French recoiled at each volley; they finally broke and fled to the rear with the riflemen in hot pursuit.[4]

16

This account bears more than a trace of Oman's musket-counting, and is in fact even more explicit. It claims that many British volleys were fired, that they were fired regularly at 15 second intervals, and that the flanks of the British line furthest from the centre of the action gradually edged forward to close the range (this was a tactic which Oman had noted in a number of British fights and which, ironically, appeared in the writings of a French theorist, Guibert). Nor is an attempted French deployment omitted. All in all this account of the action represents a textbook statement of what is today the orthodox view. It makes it clear that the French were defeated purely by British infantry fire.

The Duke of Wellington (or Wellesley), however, who effectively commanded at Vimeiro, did not see things in quite the same light. He said that the French were 'Checked and driven back only by the bayonets of that corps' (i.e. the 50th).[5] This brief comment does not, admittedly, tell us very much, and it is possible that Wellington was using the word 'bayonets' as shorthand for 'infantrymen' in the same way as it was conventional to refer to the number of cavalrymen in a squadron as so many 'sabres'. Nevertheless his phraseology at least permits us to keep an open mind on whether the French were in fact defeated by fire or by real bayonets.

Descending to the other end of the scale, we find that Rifleman Harris was a little more specific. He said that the regiment charged grandly, and 'The French, unable even to bear the sight of them, turned and fled.'[6]

This account does not mention fire any more than the Duke's, but it does make two interesting points. Harris states that the British made a charge, and that it was what they looked like which beat the enemy.

Another brief mention of the fight comes to us from General Anstruther, who commanded the next brigade on the left flank of the 50th. He must have been at some distance from the action, although he did send one of his battalions, the 3/43rd, to support it. He says little about what happened, but confirms that the British charged the French: 'The 50th Regiment, however, by a very bold attack, defeated the enemy opposed to them, taking all their guns, tumbrils, etc.'[7]

Considerably more circumstantial accounts of this little battle come to us from two officers who were actually involved in it. The first is Captain Landmann, who commanded the army's engineers and was initially posted immediately behind the 50th's position. He describes the development of the preliminary skirmishing which was done at long range by British riflemen reinforced by line companies, including some individuals from the 50th itself. The latter would bob up from behind the crest of the hill, fire their shot, and then retire to the sheltered position where the main body was waiting.

He says that the French halted to form up at long musket range from the British crestline (i.e. 2–300 metres?), and then came on in a column, heading straight for Robe's artillery battery which was just to the left of the 50th. Roundshot cut lanes through the French ranks, and cannister hit many soldiers in the front of the column. The British gunners refused to abandon their pieces and continued to fire as the French came closer:

The enemy's column was now advancing in a most gallant style, the drum by the side beating the short taps, marking the double-quick time of the pas-de-charge. I could distinctly hear the officers in the ranks exhorting their men to persevere in the attack, by the constant expressions of 'en-avant – en-avant – en-avant, mes-amis', and I could also distinguish the animated looks and gestures of the mounted officers, who, with raised swords, waving forwards, strongly manifested their impatience at the slowness of their advance, and to which they also loudly added every expression of sentiments, which they thought best calculated to urge their men to be firm in their attack and irresistible in their charge.

In this way, the enemy having very quickly approached the guns to within sixty or seventy yards, they halted, and endeavoured to deploy and form their line, under cover of the Voltigeurs. I was then by the side of Anstruther, to whom I said, 'Sir, something must be done, or the position will be carried.' The general replied, 'You are right;' and, without a moment's delay, he called out to the 43rd and 50th Regiments, as he raised his hat as one about to cheer, 'Remember, my lads, the glorious 21st of March in Egypt; this day must be another glorious 21st.' I have no doubt that this appeal had its effect.

Walker immediately advanced his gallant 50th to the crest of the hill, where he gave the words, 'Ready, present! and let every man fire when he has taken his aim.' This order was most strictly obeyed, and produced a commencement of destruction and carnage which the enemy had not anticipated. Then Walker called out, raising his drawn sword and waving it high over his head, 'Three cheers and charge, my fine fellows!' and away went this gallant regiment, huzzaing all the time of their charge down the hill, before the French recovered from their astonishment at discovering that the guns were not unprotected by infantry.

This rush forward was awfully grand; the enemy remained firm and almost motionless, until our men were within ten to twenty yards from them; then discharged a confused and ill-directed fire from some of their front ranks, for the line had not yet been formed to its full extent, and the rear were already breaking up and partially running off. The whole now turned round and started off, every man throwing away his arms and accoutrements . . . .[8]

The above account offers many points of interest. It shows that very considerable fire had been put into the French formation by skirmishers and artillery before ever the two main infantry lines saw each other. It also suggests that the firm countenance of the gunners, still without the appearance of the British infantry, was enough to persuade the French to halt, and even to deploy into line at fairly close range from their intended target. Finally, it shows that the action of the British infantry line itself consisted of a movement in three parts: first an advance to the crest of the hill, then a volley, and then an instantaneous bayonet charge with cheering, before the French had time to recover their balance. This would certainly confirm Colin's theory that French attacks were ambushed from dead ground before their deployment was completed;

but it gives no support to the view that a protracted line infantry duel (with volleys 'at regular 15 second intervals') had any part in the result. It seems to have been rather the British assault, and the effects of surprise, which were decisive.

Colonel Walker, who commanded the 50th, tells a similar story. He says the French column was:

Much shaken by the steady fire of the artillery; after a short pause behind a hedge to recover, it again continued to advance; till Lieutenant Colonel Robe, RA, no longer able to use the guns, considered them lost. Up to this time the 5th had remained at order arms, but as it was impossible, on the ground on which it stood, to contend against so superior a force, and Colonel Walker having observed that the enemy's column inclined to the left, proposed to Brigadier-General Fane to attempt to turn its flank by a wheel of the right wing. Permission for this having been obtained, this wing was thrown into echelon of companies about four paces to the left, advanced thus for a short distance, and then [was] ordered to form line to the left. The rapidity, however, of the enemy's advance, and their having already opened a confused though very hot fire from the flank of their column – though only two companies of the wings were yet formed [i.e. on the British side] – these were so nearly in contact with the bearing on the angle of the column that Colonel Walker, thinking no time was to be lost, ordered an immediate volley and charge. The result exceeded his most sanguine expectation. The angle was instantly broken, and the drivers of the three guns advanced in front, alarmed at the fire in their rear, cutting the traces of their horses, and rushing back with them, created great confusion, which by the time the three outer companies could arrive to take part in the charge, became general.[9]

Walker here refines Landmann's account at a number of points. Firstly, he does not say that the French attempted to deploy into line, but

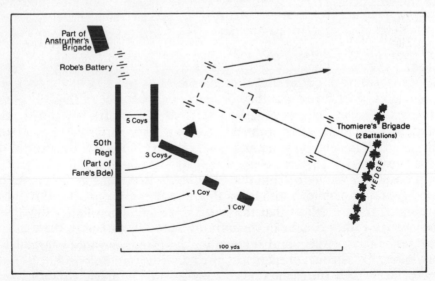

Battle of Vimeiro 1808.

continues to refer to a 'column' with a distinct 'angle' at the point of his attack. He does mention a French pause behind a hedge to regroup, which is probably Landmann's first halt at long musket range; but it is curious that he makes no mention of a later halt to deploy.

Secondly, he claims that it was his conversation with his own brigade commander, Fane, which initiated the forward movement of the British infantry. Landmann claimed it was *his* conversation with Anstruther, well to the left of the 50th, which started the movement of both the 50th and the 43rd. We can perhaps suggest that both were right, if we guess that Walker and Fane started the 50th, while Landmann and Anstruther started the supporting advance of the 43rd. The 43rd was part of Anstruther's Brigade, so this would have been the natural command responsibility.

Walker also makes it clear that he attacked the enemy in their flank rather than frontally, and that only two British companies were fully formed when the attack went in. The remaining companies of the 50th were left either in front of the French or echeloned further to the right, not yet ready to assault. If we were to follow Oman's habit of counting the muskets which could bear, we would have to conclude that this dispersion of the British considerably reduced their own firepower, while by appearing on the enemy's front and flank together they must actually have improved the French score. Walker indeed complains of a 'very hot fire' coming from the attacked flank. But we do not need to count muskets, since it is plain that the British fired only one partial volley. They defeated the French by their forward impetus and the panic retreat of the advanced French gunners who rode straight back into their own troops.

British firepower from artillery and skirmishers at long range doubtless played a significant part in this action, but sustained volleying from 'the thin red line' did not. Instead it was a number of psychological shocks, coming one after the other in quick succession, which overthrew the enemy. First there was the unexpected appearance of the 50th; then its partial volley, cheering and reckless dash forward; and finally the stampede of the French gunners. All this happened so fast that the French had no time to work out what had hit them. All they could do was run away.

Those who fondly believe that Wellington's infantry was splendidly stodgy and immobile in its battles should note well what the 50th did at Vimeiro. It had 'solidity' in the sense that it retained its coolness and performed rather complex echelon movements remarkably close to the enemy; but it was far from static. Its great strength lay in its astonishingly daring yet perfectly judged offensive stroke.

Walker's men rushed upon the enemy with the bayonet; they made what is termed a 'bayonet charge'. The fact that the enemy chose to run away before bayonets were actually crossed does not alter this fact. As it happens, Rifleman Harris suggests that some genuine bayonet fighting did take place on this occasion; but whatever truth there is in this story it was at least the British *intention* to close with the bayonet. The purpose of their musket volley was apparently the same as that of their cheering – to make a frightening noise which would enhance the effect of their attack. The British did not expect to kill the French by shooting them, but by bayonetting them. It was to the advantage of all concerned that in the event the French preferred to sacrifice their discipline, cohesion and accoutrements rather than their lives.

A study of British infantry tactics in the Napoleonic Wars reveals that this reliance upon the bayonet was far from rare. Reports of officers' words at the moment of going into action certainly stress the point. Thus at the Battle of Busaco, 1810, Wellington told General Hill: 'If they attempt this point again, Hill, you will give them a volley and charge bayonets.'[10]

Also at Busaco, Colonel Wallace of the 88th briefed his unit as follows: 'Pay attention to what I have so often told you, and when I bring you face to face with those French rascals, drive them down the hill – don't give the false touch, but push home to the muzzle!'[11]

This last expression seems to have been rather a favourite of Wallace's, since he used it again during Pakenham's decisive attack at the Battle of Salamanca, 1812, telling his men to: 'Push on to the muzzle.'[12]

General Pack, at Waterloo, briefed the 92nd Highlanders as follows: '92nd, everything has given way to your right and left and you must charge this [approaching French] column.'[13]

It would admittedly be a mistake to place too much reliance upon reported exhortations of this type, since they have no doubt been much embroidered in the telling. There are, however, many more of them in the same vein, as well as considerable evidence of other types that the British were known for their ability with the bayonet. Thus General Wilson, who was attached to the Russian army, said that the Russians were very good with the bayonet, and only one other army approached them in this – the British. A French general at Waterloo also praised the British infantry; not for its fire, but because 'It does not fear to close with its enemy with the bayonet.'[14]

In most of the Peninsular battles, and in most of the combats during the Hundred Days, we find British infantry waiting until the French came to close range, then starting a counter-advance to their own; usually with a single volley to empty their muskets, and always with plenty of cheering. In Wallace's action at Salamanca, indeed, we learn that the first cheer alone had a rather dramatic result:

The effect was electric; Foy's troops were seized with a panic, and as Wallace closed upon them, his men could distinctly remark their bearing. Their mustachioed faces, one and all, presented the same ghastly hue, a horrid family likeness throughout; and as they stood to receive the shock they were about to be assailed with, they reeled to and fro like men intoxicated.[15]

It seems that the close range counter-attack was a standard procedure with Wellington's men. It offers a good example of how an army may develop sound fighting techniques in the rough and tumble of combat which are not mentioned in its drill books. The word was apparently spread informally among the officers without being codified in writing. This, of course, is confusing for historians who have only the formal written documents to guide them, and many are those who have believed that drill books contain the key to what really happened. Drill books, however, have only a somewhat limited application in battle, and are often hopelessly out of date. (The present author's uncle, for example, was taught from his OTC drill book in the 1920s to form square against cavalry!) In Napoleonic times drill books provided only the building-blocks of tactics – the mechanical procedures for bringing each company from one position to another, for changing formation, or for maintaining alignment. The tactical use to which these methods were put was left to the local inspiration of battalion commanders and, as we have seen at Vimeiro, their brigadiers.

The British drill book during the Peninsular War was Dundas' 'Eighteen Movements', based directly upon Frederick the Great's Prussian drill. It rested on the assumption that opposing lines of infantry would form up for a duel of musketry, and that they would fight by fire. Dundas in fact specifically rejected the formation in two ranks (rather than three) partly on the grounds that it might tempt an enemy to make a bayonet attack, and so spoil the firefight. It is perhaps significant, therefore, that in practice the formation in three ranks was one of the first parts of Dundas to be abandoned. British infantry had fought in two ranks in Egypt, and formally adopted this system in 1801.[16]

Wellington's infantry did from time to time fight by fire according to the Dundas and Oman model; but this was much less common than is generally believed. One reason for the common belief, perhaps, is the coincidence that untypical firefights chanced to develop in two of the actions which have most seized the public imagination. These two famous incidents came when seven British battalions made a gallant stand against heavy odds at Albuera in 1811; and when Maitland's guard brigade defeated the French Imperial Guard at Waterloo. It may be worth our while to look at these two fights in greater detail.

Albuera is celebrated for the same reason as Balaklava; the plans of the high command went wrong from the start and had to be saved from disaster by feats of exceptional bravery and sacrifice by the rank and file. When Beresford's Anglo-Spanish army was outflanked by a heavy

French force and caught trying to change front, Colborne's brigade was sent in piecemeal to plug the gap. This unit was just developing a successful counter-charge when it was overthrown by the unexpected and simultaneous arrival of a cloudburst and a charge of French lancers. Only one of Colborne's four battalions survived the shock, which once again uncovered the army's flank. After this two new brigades, Hoghton's and Abercrombie's, had to be fed in successively over the broken ground which had already been disputed so fiercely. With a heavy enemy infantry column in front of them, artillery firing grapeshot into them overhead, and cavalry threatening their flanks, it was hardly surprising that these troops were unable to achieve a quick decision by a brisk bayonet charge. They did, apparently, edge forward gradually; but:

Unfortunately the intervention of a steep but narrow gulley rendered it impossible to reach the enemy with the bayonet, and the 29th [the leading British battalion] was directed to halt and open fire.[17]

It was in these general conditions, with or without a gully, that the famous firefight of Albuera developed. Far from being the chosen tactic of the British, it was forced upon them by circumstances. In the event the two opposed masses of infantry faced each other for about 45 minutes, exchanging shots and suffering exceptionally heavy casualties. Neither side had any offensive impetus remaining to it, and the nature of the terrain may also have made close combat difficult. The deadlock was resolved only by the arrival of new troops, especially Myers' Fusilier brigade, against the French left flank and rear. It was this fresh impetus against the exhausted units already in the fight which won the day for the British.

Albuera demonstrates that attacking troops will lose their forward motion if they are allowed to settle down to a firefight. This chanced to happen to both sides at the same time, and the result was a perfectly indecisive butchery. It was resolved only by the arrival of fresh troops who retained the ability to press forward.

Because the British at Albuera did hold their position for 45 minutes against superior numbers, it could be argued that their tactic was successful. If we remember that each of the battalions concerned lost about two-thirds of its strength, however, we may be permitted to wonder whether they could not have won the day more economically by other means. Those who suggest that the British always wanted to fight by fire would appear to be advocating a process which was excessively lengthy, inconclusive and costly. Small wonder that a briskly decisive bayonet charge was usually preferred.

Towards the end of the Battle of Waterloo Napoleon launched an attack with six battalions of his Guard, of which five battalions[18] made their assault against the low ridge running between the farms of Hougoumont and La Haie Sainte. As they advanced, these units fanned out to give themselves elbow room; but in so doing they lost the precision of their dressing, and their line became staggered. The result was that each battalion's attack appeared to be a separate action, and each one went in at a slightly different time. This has led to a great deal of confusion, since by that stage of the battle there was a lot of smoke in the air, and few observers could see very far to either side. Accounts therefore vary wildly as to when 'the Guard attack' occurred, or what became of it. Every unit which encountered one or other of the French battalions is anxious to claim the honour of having defeated the entire Imperial Guard.

The popular view of what happened is predictably the most emotionally pleasing, since it has 'the French Guard column' meeting Maitland's line of British Guards – surely a fitting final contest in a battle where the best French commander met the best British commander for the first and last time. Rising from their concealed position the two British battalions, in this account, pour in a terrific fire of musketry for minutes on end until the French eventually break under it and run away. As always, of course, most versions point out that a line formation can bring more muskets to bear than can a column, so its victory is mathematically assured. Most accounts also specify that the French made an attempt to deploy, so that the decision could be reached entirely by musketry rather than by bayonets.

Another popular touch is the decisive personal role attributed to the Duke of Wellington himself, who is supposed to have shouted 'Up Guards' to Maitland's men at the moment when the French approached to close range. Funnily enough the folk memory of this intervention has the Duke saying 'Up Guards and at 'em', which would imply an offensive purpose. Believers in the firepower theory, however, claim that he actually said 'Up Guards, make ready, fire' (Weller) – apparently following two French authorities who give 'Debout Gardes, et visez bien' (Mauduit) and 'Debout Gardes et visez juste' (Charras). The greatest student of Waterloo, Siborne, variously quotes either the simple 'Up Guards' on its own, or the more aggressive 'Up and Charge!'; but it is probably the opinion of Major Saltoun, an eyewitness, which is nearest the mark. He says Wellington said nothing at all, and that the tactical decisions were all taken by Maitland and his subordinate officers.[19]

However this may be, there is a pretty fair agreement in the sources that Maitland's men did defeat a French Guard column by a lengthy exchange of musketry; that the enemy did attempt to deploy into line; and that he eventually broke before the British started to move forward in pursuit. For all our reservations, therefore, we must accept this as a rare bona fide case of a successful firefight. Where British line fire at

Albuera had failed to rout the French, British Guard fire at Waterloo apparently succeeded.

There is nevertheless a lot more to the story than this, since Maitland was facing only two out of the five French battalions. What happened to the other three? It is difficult to say with certainty, because eyewitnesses were often themselves unclear. It seems, however, that all three of the remaining French battalions were repulsed by bayonet charges, and did not attempt to deploy.

To the right of Maitland Major-General Adam faced the advancing French with the 1st Battalion of the 52nd Regiment supported by some riflemen. He afterwards explained that 'It was not judged expedient to receive this attack, but to move forward the brigade and assail the enemy, instead of waiting to be assailed.[20]

There followed an attack which must have been almost identical to Walker's at Vimeiro. The 52nd swung forward its right flank, refused its left, and charged in with cheers and a volley. The flank of the enemy column put out a damaging fire while the manoeuvre was under way; but as soon as the British came to the assault the whole mass turned tail and ran.

On Maitland's left Sir Colin Halkett's Brigade had received a terrible battering during the campaign, and was reduced from four battalions to two weakened composite ones – the 73/30th, and the 69/33rd. The latter unit was apparently involved in the outskirts of Maitland's battle at one point, yet it later gave a volley and made a charge in cooperation with the 73/30th against another enemy column. This French unit ran away with 'inexplicable' suddenness, and confirmed Major Macready of the 30th in his opinion that:

All firing beyond one volley in a case where you must charge, seems only to cause a useless interchange of casualties, besides endangering the steadiness of a charge to be undertaken in the midst of a sustained file fire, when a word of command is hard to hear.[21]

After their successful attack, however, there was a certain amount of confusion within Halkett's Brigade, as there was also among the Brunswickers on their left flank. The inclination to retire was nevertheless eventually overcome, and a line was consolidated. Desultory long range shots were then exchanged for a while until a new French column attack appeared to be in preparation. This threat never actually materialised, and was finally chased away by the combined effects of an advance by the Dutch-Belgians from behind Halkett, 'waving their shakoes on the ends of their bayonets'; by the forward movement of allied cavalry; and by the spread of despondency among the French following the defeat of their other attacks.

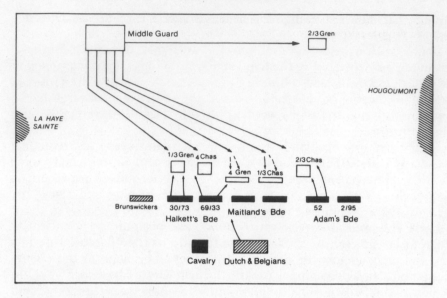

The Guard attack at Waterloo.

This group of actions illustrates quite neatly the variations in technique which were available to British infantry, and it confirms our impression that the bayonet counter-attack loomed large in their thinking. Macready's comments are particularly interesting, since they emphasise one of the great weaknesses of sustained volley fire. This was its tendency to degenerate into a ragged 'fire at will' after the first command volley. When this happened the fire became very hard to stop, and words of command were drowned by the din. At the Battle of New Orleans early in 1815 General Pakenham was killed while unsuccessfully trying to make his storming parties cease fire and move forward. His fall, in fact, served only as a signal for the redoubling of the fire. At the assault on Buenos Aires in 1807, equally, some of the attacking forces disobeyed their orders to move forward without loading, and halted to fire. 'From that moment their advance was at an end', as one report put it.

Major Patterson, who served with the 50th in the Peninsula, made a similar point:

Inexperienced officers have repeatedly given orders to commence a fire, without either judgment or consideration as to whether or not it was the proper time to open a fusilade. This was their fault and not the men's. However, the mischief to which it tended was, that after the first command was given, the soldiers of themselves, taking out a sort of carte blanche, blazed away, in the most independent manner, in all directions, until at length the utmost skill and energy of the most active officers was baffled, in their efforts

26

to controul (sic) them; and when the ammunition was most required they found it was expended to little purpose, beyond that of raising noise and smoke.[22]

A number of British officers did come out of the Napoleonic Wars with the idea that a formal contest of musketry, as recommended by Dundas, was the best tactic; but it would seem that rather more took the opposite view, and put their faith in a quick volley followed by loud cheering and a bayonet charge. Believers in this doctrine were not deterred by the argument that only firepower killed people in battle, since they were less concerned with killing the enemy than with destroying the cohesion of his units and chasing him away from the places where he might be dangerous.

There is certainly plenty of evidence to suggest that bayonets caused very few casualties in battle compared with the great majority of wounds which were inflicted by shot and shell. There are also remarkably few accounts of bayonets actually being crossed by two forces which were facing each other. If this happened at all it would normally be in circumstances where neither side could easily escape, such as in buildings, fortifications, sunken lanes or in a night surprise. There are admittedly several claims that genuine bayonet fighting took place in the open, as at Maida, 1806; Rolica and Vimeiro, 1808; Busaco, 1810; Barossa, 1811; or at Maya and Roncesvalles, 1813. In none of these cases, however, were very many troops involved, or, indeed, are the accounts themselves particularly conclusive. It seems that in the close fighting at Maya, for example, some of the troops preferred to throw rocks from a distance rather than close for true bayonet fencing.

We can therefore quickly dismiss the romantic notion that Napoleonic soldiers in open warfare habitually '. . . battered each other with musket butts and bare hands, they stabbed and clawed . . .' (this from a description of the firefight at Albuera which appeared as recently as 1973!). If they came close enough to physically stab their enemies it was usually in pursuit of a beaten force, or to finish off wounded enemy soldiers. One officer of Pack's Brigade at Waterloo did not believe that these latter practices took place at all, although his other comments ring true enough. He said that:

With regard to any 'bayonet conflict', I saw none. We appeared to charge, and disperse, and make a road through the [French] columns – the usual result of the British charge. This accounts for the absence of bayonet wounds, on which Colonel Mitchell builds his theory of the uselessness of that weapon. The weaker body generally gives way: after which what British soldier would bayonet a flying enemy?[23]

One answer to his final question is that at Busaco the 52nd Regiment turned an enemy column off the top of the ridge, and then 'kept firing and bayonetting them until we reached the bottom'.[24] Something very similar must have happened in many other battles, although it is not usually described in much detail by participants.

If the success of a bayonet attack could scarcely be measured in terms of the casualties inflicted on the enemy, however, there was at least one solid index of victory which had more than a passing interest for the Napoleonic soldier. This can be summed up in the word 'booty'. When a body of infantry turned in flight each man would normally hasten his departure by discarding his – highly plunderable – heavy equipment. Thus when Pack's Brigade at Waterloo was at one stage brought face to face with a French column:

. . . some British officer called 'Charge! Charge!' (he was directly knocked over with the word in his mouth), on which the head of the French column got confused, threw down its arms, accoutrements, and knapsacks, and surrendered.[25]

Many other accounts speak of lines of knapsacks being left on the ground at the spot where a unit had turned tail. These items were an extremely heavy encumberance to movement, and attacking troops might sometimes leave them behind deliberately to facilitate their movements – as Kempt's men did at Maida, and as the 43rd and 52nd Regiments did on the Bidassoa. Normally, however, they were far too valuable for soldiers to part with except in the direst emergency. They contained food, blankets, spare woollens, and a host of other personal effects which could make life on campaign almost bearable. When a unit involuntarily abandoned its knapsacks, therefore, it was a sure sign that it had been truly beaten.

The champions of firepower are right to say that musketry was more lethal than bayonets; but this is not to say that it was more decisive. The most damaging form of musketry, after all, was aimed skirmish fire: a type of action which was specifically designed for attrition and not for reaching a decision. Troops formed into close order found it much harder to aim effectively than dispersed skirmishers, and it was only rarely that they were expected to aim at all. Volley fire was thus an extremely wasteful process. Most Napoleonic authorities calculated that only something like 0.3 per cent of all musket balls fired would hit anyone in battle. (See discussion in the additional section at the end of this chapter.)

Yet the myth of sustained British volley fire persists, even down to some very anachronistic and inflated claims for the Brown Bess musket itself. Because it fired a slightly heavier ball than the French 1777 musket, some modern commentators have assumed that it must therefore have been a greatly superior weapon. The officer we have quoted from Pack's Brigade, however, said of the French that:

Their fine, long, light firelocks, with a small bore, are more efficient for skirmishing than our abominably clumsy machine. . . .

He went on to say that the British muskets were:

. . . of bad quality; soldiers might be seen creeping about to get hold of the firelocks of the killed and wounded, to try if the locks were better than theirs, and dashing the worst to the ground as if in a rage with it. I believe this was quite common throughout the war in Spain.[26]

The truth of the matter is that both sides believed their enemy had superior muskets, just as in Vietnam both sides believed the enemy had the better assault rifle. It is only natural, after all, that one should imagine a weapon is more dangerous if it is carried by an enemy rather than by a friend.

The real secret of the British volley was not that it was delivered particularly well or accurately, but rather that it could be delivered at all at such close range: almost at bayonet range, in fact. Waiting without firing until the enemy came close enough to charge must have been a nerve-racking business which tested the coolness and discipline of the troops to the limit. Having delivered the volley, it took yet more steadiness not to reload, but to launch immediately into the assault. Paradoxically, therefore, we can say that it was the British ability *not* to fire at the wrong moments, rather than their skill in musketry, which enabled them to get the best from their weapons. Other armies fired more and at longer ranges, but found the final result less satisfactory. Colonel Mitchell, for example, said the French were always ready to '. . . faire le coup de fusil, as they termed it, for hours and days together. But who ever saw them await a bayonet onset?'[27]

But the palm must go to the Spanish at Talavera. Noticing a few enemy horsemen at about a kilometre's range, the entire Spanish line opened fire with its muskets and then all ran away in panic. They were 'frightened only by the noise of their own fire', as Wellington put it. This sort of thing represents the diametrical opposite of what the British could achieve in their controlled counter-charges.

We have now analysed Oman's theory of British firepower at some length, and found that contrary to expectation it was apparently the French, if anyone, who preferred firing to making a bayonet charge 'while the English invariably seek to close with their enemy'.[28] Insofar as Oman believed the British fought primarily by volleys he was certainly wrong; but we must ask if he was also wrong to deny Colin's theory, and say that the French did not want to develop a firing line of their own in these battles. It is time now for us to turn to the French side of the hill.

We are much less well served with French eyewitness accounts of tactical actions than with British. The bulk of the French forces, and

29

especially the élite units which tend to produce most memoirs, fought on the 'Eastern Front'. They came to Spain only for the campaign of 1808–9, when most of the opposing armies were Spanish; and they met the British only in 1815 at Quatre Bras and Waterloo. Remarkably few French generals fought more than one or two battles against the British, so they perhaps thought of those experiences as something of a flash in the pan. Most accounts of their tactical methods gloss hastily over their inglorious performance against Wellington's infantry, and concentrate on the more successful and important battles fought against the vast armies of Eastern Europe.

Another factor which restricted the debate on minor tactics in the French armies was that the very idea of tactics was conceived on a much more grandiose scale than in the small and closely-knit British force. For a French commander the important thing was to manoeuvre large formations – whole divisions and army corps – into contact with the enemy, whereas the British could afford to give more individual attention to the action of brigades and even single battalions. This meant that the French were thinking of tactics at a higher level than the British, and perhaps tended to be rather slap-dash about the details.

We cannot avoid the impression that there was no French 'standard procedure' for minor tactics comparable to the informal British agreement on close range counter-charging. Every French general appears to have experimented with drills and formations of his own, and these were never standardised. Napoleon had intended to produce a manual for tactical guidance around 1812; but this resolution unfortunately became yet another casualty of his Russian campaign. As with the British the only drill book remained based on eighteenth century practice, and was honoured as much in the breach as in the observation.

Some commentators have claimed that the French lack of standardisation in minor tactics was actually a strength, and that it demonstrated the flexibility and professional skill of their generals. The bewildering variety of their formations is hailed as proof of innovative or original thinking, and Napoleon himself is particularly praised for his use of the 'ordre mixte'. This was a mixture of columns and lines designed to get the best out of both; but one suspects that in practice it more often managed to produce the worst aspects of both. It was rarely seen on the battlefield, for although Napoleon frequently recommended it to his generals they tended to abandon it hastily as soon as he was out of sight.

Praise for these 'experiments' in tactics really stems from Bonapartist hero-worship, and has on occasion been linked to such astonishing statements as 'Wellington was no match for the Emperor as Waterloo proved'.[29] If we look at the matter more objectively, however, we can see the French tacticians struggling to evolve methods for manoeuvring formations which were much larger than those envisaged by the 1791 drill book. They actually succeeded in this as far as bringing troops into musket range was concerned, and that was no mean achievement. They failed, however, when it came to the moment of contact itself.

This failure to resolve the problem was noted by the British Colonel Mitchell, writing in 1833. He was amazed that:

. . . French writers have actually discussed the point, whether columns were intended to fight, or only to move, so that it seems they do not yet know the object of the very formation with which they all but conquered continental Europe.[30]

A later French authority, Colonel Ardant du Picq, came to the same conclusion. 'The cavalry has definite tactics;' he wrote, 'essentially it knows how it fights; the infantry does not.'[31]

The French infantry did at least pride itself on the particularly high proportion of officers and NCOs in their battalions, as well as upon the cadres of veterans who set an example to the raw recruits and gave a solid backbone to each unit. This system of leadership was called 'surveillance', whereby there were always plenty of old hands to watch over the new arrivals. It was very successful in maintaining the resilience of units which had to be renewed only too often as the wars dragged on; and it quickly inducted fresh troops into the arts of foraging and survival on campaign. Where it was apparently less successful, however, was in the realm of tactics.

Each French regiment usually sent a number of its battalions into action together, so that the regimental commander had several battalion commanders of the rank of major under his hand. This compares unfavourably with the British practice of fielding only one battalion from each regiment, directly under the regimental commander himself. The French system was thus over-officered at company level, but relatively under-officered at the level of battalion command. Regimental commanders had too many sub-units, and they were tempted to keep them too closely bunched together in order to maintain personal control. Each sub-unit was also rather too small, since following the decree of 18th February 1808 there were only six companies in each battalion as compared with the British ten. When the wars were over this was a point which attracted a great deal of criticism from French military writers.

Another source of trouble, as both Jomini and Oman pointed out, was that in most of the French battles against continental enemies there had been little call for massed infantry to do very much real fighting at all. Their mere appearance within musket range of the enemy had been sufficient to decide the issue, or if not, there had been adequate supports at hand to bail them out. Massed action by cavalry and artillery was far more a feature of these battles than it was in the Peninsula, and Sergeant Bourgogne said of Borodino, for example, that 'This, like all our great battles, was won by the artillery.'[32] It was noticeable that in Napoleon's only battle against Wellington, at Waterloo, it was the artillery which decided the timing of the whole action, and cavalry attacks which took up a good half of it.

The superiority enjoyed by the French in their continental battles must have led them to regard infantry attacks as somewhat expendable. They would start by softening up the enemy with artillery and skirmish fire, and then send in a first wave of massed infantry. If this attack failed it would not usually be sent back pell mell, as it would have been by a British counter-charge. Instead, it would be allowed to remain at musket range of the enemy and act as a thickening for the skirmish screen which was already in place. The pressure on the enemy would be increased, even though the attack itself had technically been defeated. The second wave could then be sent in, and if that also failed then the pressure against the enemy would nevertheless have increased yet again, and so on. General Brun de Villeret gives us a description of this process in his 'Cahiers'. At the Battle of Bautzen in 1813 he was ordered to assault a Russian hill position with two battalions from his brigade. He did so:

. . . but the Russians fought back, and quite a lively fusillade started. I then decided that I should support them with two other battalions, and I marched at their head. At last, seeing the affair was getting serious, I advanced the last two [battalions] and the position was carried.[33]

The defeat of an infantry attack was therefore seen as a temporary rather than a serious reverse, and no one bothered to analyse in detail precisely what had produced it. When defeated rather more conclusively by the British, the French apparently fell back on the same familiar line of reasoning, and shrugged it off. They failed to realise that in this case the defeat tended to strengthen rather than weaken the moral balance in favour of the enemy, or that the habitual support of massed cavalry and artillery was in all probability lacking. French accounts of Peninsular battles against the British are remarkable for their complacency.

In the light of all the above factors it is hardly surprising that French Peninsular tactics were neither intrinsically strong nor well reported in the documents. They did not seem important to very many French generals, and they were certainly not glorious. Extremely heavy columns appear to have been thrown into battle rather carelessly, and deployments attempted as an afterthought if they were attempted at all. Thus far we must agree with Oman's version rather than Colin's, although we must attach a number of caveats to this finding.

In the first place we must be very careful when we use British accounts of what they were up against. There seems to have been a natural tendency for British observers to overestimate the size of columns which were attacking them. Where we can compare actual numbers with

British reports we often find only half or a quarter as many French troops as were claimed. At Vimeiro, for example, both Landmann and Walker claimed that the 50th was attacked by 5,000 men, although the true figure seems to have been nearer 1,200. At Waterloo, again, many British reports speak of many battalions in each Guard column, whereas we know from the French side that each one can have contained no more than two battalions. There seems to be a consistency in these overestimates which suggests that the French columns were really a good deal lighter than often assumed. It is true that they were rarely composed of single battalions as their best tacticians would have liked; but at least they were seldom as monstrous as British propaganda would have us believe.

We must also beware of the British accounts when they speak of attempted French deployments, since it is seldom made very clear exactly what was involved. Few British reports specify whether these deployments were ordered by officers or occurred spontaneously; a point of cardinal importance if we would fathom French intentions. In fact both types are mentioned in different circumstances; but once again there is an annoying reticence over the precise timing of the deployments concerned. We cannot often judge whether the French started to form a line before or after they had run into trouble. This might have told us whether the deployments had been planned from the start or were simply an emergency response to an unexpected development. It is true that a number of British accounts speak of the French standing still as if dazed or shocked before attempting to form line, so our impression tends to be that deployments were in fact improvised and hasty. This is confirmed by the apparent rarity of deployments which were fully completed before their rout. We have, nevertheless, very little conclusive evidence to work from.

There is the further possibility that many deployments may have consisted, in effect, of an instinctive flight to the flanks. Men in the centre of a column which has suddenly stopped will be prevented from advancing by the immobilised front ranks, and prevented from retreating by the continuing advance of the men behind. They will have nowhere to go, assuming that they still wish to move, except to the flanks. It seems likely that this mechanism would create the impression of attempted deployment in the eyes of someone watching from the front.

This question of French deployment for the attack also sounds another warning bell in connection with Oman's work, since he based a good part of this theory upon the Battle of Maida, in Calabria. Although this was not strictly a Peninsular battle it seemed to fit perfectly into his explanation for the combat of the column against the line. When he first drew up his position in 1908, he used Maida as an example and counted

33

the muskets which could bear from the British line as compared with those of the French column. Unfortunately, however, by 1910 he had discovered that at Maida the French were in fact deployed in line, so his original mathematics were shown to be spurious. Worse; the fact that the French had been in line also gave support to Colin's claim that the line attack was their normal drill. It seemed that the very foundation of Oman's argument had been removed. Apparently untroubled by this finding, however, he stuck to his original calculations of firepower, and in 1929 he even reprinted the first article as it stood.[34]

This episode confirms our distrust of Oman's confident musket-counting; but on the other hand it does little to help Colin. By chance Maida is one of the few battles for which we have a clear report of what the French had intended, and it transpires that although they certainly deployed, they were also under orders to attack without firing. The French were in line not to develop their firepower at all; but to make a bayonet attack. They might well have succeeded, had not the British forestalled them by one of their timely counter-charges. It was this which over-awed the French, who 'saw that they had exchanged roles with the English, and that they were no longer the assailants'.[35]

Colin's argument that the French intended to deploy and attack by fire is based upon very shaky evidence. He makes the good point that line attacks were made in several French battles outside the Peninsula,[36] but even he admits that few cases can be found from the Peninsula itself. A number of after-action reports by French generals, however, can be seen as supporting Colin's position. Thus we hear that at Vimeiro Solignac's grenadier regiment was defeated because, '. . . led with too much ardour, [it] . . . did not have time to deploy, and was overthrown'; while in the same battle Kellerman's grenadiers 'didn't even have time to complete their development'. After Albuera Soult complained that his men would have carried the day if their front line 'had deployed and presented the enemy with an equal frontage; but, there was hesitation . . .'[37]

These reports prove that at least some senior French commanders recognised with hindsight that their columns ought to have deployed; but they gloss over the question of why the attempted deployments failed. The explanations offered seem rather inadequate. 'Excessive ardour', 'insufficient time', and 'hesitation' form rather feeble, not to mention contradictory, excuses. We are left wondering whether or not the tactical controllers of these defeated units had seriously wished to deploy at all. The impression is that they wanted to do so only after their original attacks in column had been halted. Against an opponent who did not make counter-charges their deployments might then have succeeded; but against British mobility and opportunism they were already too late.

There was thus a theoretical acceptance of deployment in the French army, originating no doubt from the 1791 drill book. There seems, however, to have been little practical effort to do anything about it in time to be useful. One reason for this was that column attacks had a theoretical justification of their own in the drill book, in that instructions for forming a battalion 'attack column' had been added as an afterthought to the complex manoeuvres of the line. A number of theorists had also advocated column attacks while some who had not, such as Ney, seem to have used them, nevertheless, in battle. The argument that column attacks had often worked well against other enemies remained seductive, not least in the Peninsula itself where it had achieved notable success against the Spanish. It would therefore have been surprising if the French had not tried to march on in column when they came against Wellington's infantry.

All French authorities were agreed that the column was the best formation for movement, especially over difficult terrain. In the Peninsula the terrain was often broken and hilly, and on some battlefields such as Busaco or the Pyrenean passes it was well-nigh impossible. To bring their men into musket range, therefore, French generals would naturally have employed the column, and it was only on arrival at this range that they would have been faced with the choice of whether or not to deploy. If they had deployed they might have gained certain theoretical advantages; but would have had to sacrifice their forward movement. It must have seemed only too tempting to continue their impetus in the formation in which they found themselves, and only too obviously dangerous to halt for manoeuvres. From first principles it is therefore easy to see why they usually continued forward in column.

The need to continue the forward impetus also dictated their attitude towards firing in the attack. The main rationale behind deployment was that it brought the maximum number of muskets to bear; but as we have already seen on the British side, opening a fusillade brought with it a high risk of losing control, cohesion, and especially forward movement. In an important way, therefore, an attack with fire was a contradiction in terms. Reynier had seen this at Maida when he ordered the attack to be made without fire, and there is evidence from other French battles that the same reasoning applied elsewhere.

Jomini, for example, said that 'During the recent wars Russian, French and Prussian columns have frequently been seen to carry positions with shouldered arms, and without firing a shot; this represents the triumph of the impetus and morale effect which is produced.'[38] At Montmirail, 1814, Marshal Ney ordered the guard to charge the enemy after shaking the priming powder out of their muskets so they could not fire; while in a skirmish after the Battle of the Katzbach in 1813 a junior officer, Martin, noted that conscripts who had been ordered to attack without firing disobeyed instructions and came to a halt short of their objective.[39]

Perhaps the most often quoted description of a French attack was written by Marshal Bugeaud and first used by General Trochu in his work on the French Army of 1867. Bugeaud, however, probably caught sight of the British only at the Battle of Castalla in 1813, and even there his unit was not involved in the fighting. For his apparently 'eyewitness' description of the typical Peninsular combat he relied heavily upon an account of Talavera by General Chambray which appeared in 1824, and it is this which we reproduce here:

The French charged with shouldered arms as was their custom. When they arrived at short range, and the English line remained motionless, some hesitation was seen in the march. The officers and NCOs shouted at the soldiers, 'Forward; March; don't fire'. Some even cried, 'They're surrendering'. The forward movement was therefore resumed; but it was not until extremely close range of the English line that the latter started a two rank fire which carried destruction into the heart of the French line, stopped its movement, and produced some disorder. While the officers shouted to the soldiers 'Forward; Don't open fire' (although firing set in nevertheless), the English suddenly stopped their own fire and charged with the bayonet. Everything was favourable to them; orderliness, impetus, and the resolution to fight with the bayonet. Among the French, on the other hand, there was no longer any impetus, but disorder and the surprise caused by the enemy's unexpected resolve: flight was inevitable.

This account is probably the best surviving summary of what must have happened in most actions between the French and Wellington's infantry. It shows that the French aim, whatever their formation, was not to open fire but to press on with the bayonet. The fact that muskets were 'shouldered' is even noted as customary practice in such attacks. Nor is the British fire singled out as particularly decisive. Admittedly it managed to halt the attack for a time; but so had the initial sight of the British steadiness. The fire is said to have caused nothing worse than 'some disorder', whereas the really decisive shock which routed the French was unquestionably the bayonet charge and the unexpected British resolve.

Colin's school can derive very little satisfaction from this passage, since it is clear that the use of fire was very far from the minds of French tacticians. The suggestion of a British ambush is also lacking, at least in the physical sense. The British did not pop up from behind a fold in the ground at the last moment, but were obviously visible for most of the time. It may be true that a physical ambush had genuinely been achieved by both Walker at Vimeiro and Maitland at Waterloo; but in Chambray's description it seems clear that the only ambush was a purely psychological one. The French were astonished by the remarkable coolness and aggressive intention of the enemy they were attacking.

This appears to be the key to the whole matter. It was the relative balance of morale and steadiness which decided the victor in an infantry fight, and not the balance of firepower. In their short sharp countercharge the British were able to retain most of their orderly bearing; whereas the French were caught at the end of a long and discouraging approach march, suffering from too many cheer-leaders at too early a stage. Because the British had refused to be intimidated by this display, the burden of doubt was inexorably transferred to the French. Their confidence seeped away to the point where it disappeared completely in face of the smallest threatening gesture.

It is highly significant that Chambray makes no mention at all of the French attack formation in his account of Talavera. He describes the important moral factors in some detail; but he seems to dismiss the question of column or line as devoid of relevance. This finally puts the bickering between Oman and Colin in its place, since we can now see not only that both were quite wrong to stress the role of musketry, but also that neither column nor line offered any particular advantage to an attacker. Because musketry was less than decisive, the line had no secret gifts to bestow. The column in turn was considerably less dangerous to its members than is often claimed. At Maida we have seen that a French attack in line suffered from precisely the same moral defeat which was later to beset column attacks in the Peninsula. As a footnote we might add that at Maya in 1813 a French column apparently defeated a British line by its fire. From this it is possible to see that the French problem was not therefore one of formation at all; it was rather a matter of the steadiness of their men at the decisive range. Beside this overwhelmingly important factor the mathematics of musketry appear to have been very trivial indeed.

When we contemplate the struggles for no-man's-land in the Peninsular and Waterloo campaigns we are forced to conclude that it was the relative quality of the opposed infantry which decided the outcome, rather than any technical considerations of formation or firepower. When French officers simply admitted that the British troops were better soldiers they actually came nearer to the heart of the matter than those who would count muskets, measure bullet sizes, or calculate deployments. One of the French officers captured at Vittoria in 1813 really summed it up when he remarked to Wellington that 'Le fait est, monseigneur, que vous avez une armée, mais nous sommes un bordel ambulant.'[41]

Some Spurious Qualities Claimed for Rifles and Fieldworks

If we say that a column formation was just as good or bad as a line, and a British musket just as good or bad as a French one, then we are undermining a considerable portion of received wisdom about Napoleonic tactics. We have not gone the whole way to total iconoclasm, however, since many commentators believe that Napoleonic troops could still win decisive advantages by exploiting two other items of 'hardware', namely the rifle and the spade. Let us now investigate these claims a little more fully, with particular reference to the ranges at which combat might be decided.

It has often been alleged that the Napoleonic smoothbore flintlock musket was a dreadfully short-ranged weapon, and that a person standing more than 150 yards from a marksman would have to be desperately unlucky to be hit, provided that he had been selected as the target.[42] It therefore seems to have been something of a rule of thumb that 'effective' combat range for troops in close order was about 50 yards, almost regardless of the type of ammunition they were firing. Whether this ammunition was the ballistically relatively efficient single bullets, Bugeaud's preferred doubled bullets ('buck 'n' ball'), or small buckshot, the general result seems to have been much the same. Numerous cases have also been cited of whole battalions firing at a hundred yards, yet inflicting only two or three casualties upon the enemy.[43] Napoleonic authorities certainly seem to have believed that a formed unit was safe enough from musketry at any ranges further than about 150–200 yards.

In Jac Weller's extensive 1954 tests this effect was explained by the wild inaccuracy of the weapon, which produced hit groupings over three feet in diameter at 100 yards, and no accuracy whatsoever beyond 200 yards.[44] Figures for ammunition consumption in battle tell a similar story. At Vittoria it was estimated that only one musket shot in 800 hit an enemy soldier, after the effect of artillery had been taken into account, while during one patrol at the Cape in 1851 some 3,200 shots were needed for each hit.[45]

Against all this, however, there is an opposite body of evidence to suggest that the smoothbore musket was somewhat better than its many critics have maintained. The French 1791 manual intended no satire when it gave a ballistic maximum range for the 1777 musket of 900 metres,[46] while in range tests, far from the emotions and stresses of combat, some very impressive scores are recorded from the Napoleonic period itself. Frequently 20% of shots hit an (admittedly large) target at 300 metres and 40% at 150 metres.[47] In routine regimental shoots during the July Monarchy the average was rather lower, standing at 14.3% hits for ranges between 100 and 300 metres; but this still indicates one hit out of every seven shots fired.[48] Actually the best recorded Napoleonic battle result does approach this level of accuracy quite closely, since a couple of British battalions at Maida used only 8.7 rounds to score each hit on the French, albeit at ranges of between 115 and 30 yards.[49] The secret here, however, seems to lie in the closeness of the range and the discipline of the firing troops – rather than in the inherent qualities of the weapons.

Normally the combat results of Napoleonic musketry fell much nearer to the Vittoria score than to that of Maida.

Wellington's infantry certainly possessed the discipline and coolness needed to maximise the effect of their weapons by holding their fire until particularly close range. A study of some nineteen British firefights for which musketry range is specifically mentioned[50] shows only four cases (21%) where fire was opened at 100 yards or more, but nine (47%) in which the combat eventually closed to twenty yards or less. The average range for opening fire (from 16 mentions) works out at 75.5 yards, and that for closing fire (from 17 cases) is 30.4 yards. The overall average range of British firefights, taking a mean between the start and finish of each engagement, thus seems to work out at 64.2 yards. It is the average for closing fire, however, which is probably the most significant: little more than the length of a cricket pitch!

It remains an open question whether the infantry of other nations could habitually hold their fire until such close ranges, especially after the passing of most of the mature, intensively trained armies by around 1808. Whether it is a matter of the *Grande Armée* of the Boulogne camp, or of the Prussian army of Jena, we have a general impression that these professional armies had been better at the minutiæ of battalion drill than were their immediate successors.[51] Personally, I am persuaded that the emphasis thereafter moved more to skirmish fire at longer range – but admittedly we have no solid data base to prove it. No specific figures or analyses have ever, to my knowledge, been assembled to test this point, and one suspects that they would be very much more difficult to find, even for the French, than they are for the British. If we *did* have such details, and if we *could* make an attempt at generalising the firing ranges for the whole of the period between 1808 and 1815, it would surely be surprising if the overall average were not considerably longer than for the British during the same period.

This is not to say that all long-range fire was always necessarily inaccurate. There was at this time, for example, a certain growth in the idea that soldiers should be encouraged to aim their pieces, rather than just level them in the general direction of the enemy.[52] This in turn was often linked to the idea of training specialist light infantry, and some proud regimental traditions were created as a result. Nevertheless, it is worth remembering that Napoleonic 'Light Infantry' was all too often indistinguishable in practice from line infantry,[53] and much nonsense has been written about the special qualities supposedly bestowed by a 'light' designation. We must therefore tread warily, and distinguish carefully between two very distinct species: frivolous light infantry who were not given any special training, and serious light infantry who were.

On one hand there was a mass of ordinary soldiers who happened to be called 'light infantry', 'voltigeurs', 'chasseurs', 'tirailleurs', and so on, merely by a terminological accident and regardless of any particular ability or experience. Such troops might be reinforced in their light identity only by some tailor's flourish in the cut of the jacket, or by an unconventionally coloured plume. In the French army, where most Napoleonic light regiments were of this type, there was no proper training manual for skirmishing before 1831, unless units chanced to receive privately produced drill books from their individual colonels.[54] All infantry, both light and line, often fought as skirmishers; but usually following merely their own inspirations.

Perhaps the most extreme case of a 'pseudo-light' formation was the Second Imperial Guard Voltigeur Division of February–March 1814. Three-quarters of its men had been civilians only ten days before the Division was formed and marched out to war, which was itself only twenty days before it had to bear the brunt of some heavy fighting at Craonne, suffering around 50% casualties. The men's personal equipment was always very far from complete during this campaign, and their formal tactical training must have been as close to zero as to make no difference.[55]

A number of France's opponents, on the other hand, did maintain small but specialist light infantry forces that were genuinely manned by experts. Such forces would typically enjoy high *esprit de corps* and a strong awareness of the true functions of light infantry – although in combat they tended to be used in close order as often as they were set to skirmishing.[56] We may further suggest that their forte lay in skilled but auxiliary outpost, scouting or rearguard work, or in preparatory, harrassment shooting before the main set-piece clashes of large bodies, rather than in the exchange of musketry between more or less densely formed lines. Such troops doubtless made an invaluable contribution to every army they supported – but they probably did not often change the overall range or rate of killing within the major firefights by a very great margin.

Those light infantry regiments which were sufficiently specialised to carry rifles could certainly achieve superior results to those armed with the smoothbore, and at longer ranges. Weller found that rifles firing a spherical ball could keep a grouping of around twelve inches at 100 yards, although accuracy at 400 yards was described as either 'poor' or 'very poor'.[57] Ezekiel Baker, the designer of the British rifle used in the Peninsula, felt that he could not guarantee accuracy at greater ranges than 200 yards, even in target practice, although a proportion of hits could still be expected at 300 yards.[58]

Nevertheless, this means that the true advantage with a rifle was really little more than 100 yards' extra reach over that of a smoothbore, which might sometimes be reduced still further by the growing interest in accurate fire that was potentially increasing the efficacy of ordinary muskets. In any case, modern experience in Korea and Vietnam has shown that there are normally very few combat situations in which riflemen have an opportunity to fire effectively at more than 200 yards, regardless of the ballistic properties of their weapons. Beyond that range the battle tends to pass to heavier, crew-served weapons – or artillery, in Napoleonic terms – if there are any targets visible at all.[59]

To compound the problem, Napoleonic riflemen always had to pay a heavy price for their slender margin of extra range, in terms of convenience and speed of firing. The need to use a patched ball to minimise 'windage' made for difficulties in ramming, and often necessitated the assistance of a mallet. Baker was also at pains to stress the great care and deliberation that was essential to loading, estimating range and aiming, which all added still further to the slowness of fire. It would take a rifleman at least three times as long to fire each round as it did a musketeer, leaving him exceptionally vulnerable to massed attacks unless he was well protected by terrain or fieldworks.[60]

In summary, it seems fair to say that there were no genuinely long-range Napoleonic weapons apart from the artillery, although the latter could certainly exercise a powerful effect up to about a kilometre, and the French Villantroys coastal mortar was claimed to be reasonably accurate against shipping as far afield as six kilometres. In common with musketry, however, artillery was generally decisive only at short range, where grape and cannister could be used. A formed line would perhaps not choose to present itself within long artillery range of an enemy for any length of time if it could be avoided; but if it was unavoidable, such an ordeal was usually considered more or less bearable. Wellington deliberately accepted this penalty for Beresford's flank march around the enemy redoubts at Toulouse, and the French Imperial Guard made a celebrated stand under prolonged cannonading at Aspern-Essling.[61] By contrast to such operations, however, it was generally understood that a much more fearsome fire could be expected within a zone 200 yards or less from an enemy's position. When they arrived within this zone it was normal for soldiers to hope for a very rapid resolution of the combat.

A walking man can cover 200 yards in two or three minutes, and this should therefore in theory have been the total time taken up by most infantry clashes. An attacker had to make up his mind to cross that distance as rapidly as possible, in order to chase off the enemy before the latter could fire many shots. With luck, during two minutes a defending rifleman might fire only once, a cannon twice or thrice, and a musketeer perhaps scarcely more. Apart from a few novelties, like the repeating Austrian airgun, a defender's weapons normally had a low rate of fire. Even when the defender was closely packed, as he often was,[62] this made a relatively low total volume of fire – at least in contrast to massed fire from either the medieval longbow or the late nineteenth-century

41

breechloading rifle. It was not very much to keep an attacker at bay, but it was the best that was available in Napoleonic times.

To further downgrade a defender's fire, the weaponry of the day was notoriously prone to misfires, and the loading procedure was so complex that all but the steadiest troops were quite liable to fumble it in a crisis.[63] Under conditions of combat stress the available fire might well lack the accuracy and density needed to stop a determined rush, allowing a defender to be swept away. From his point of view, therefore, it was important to find some means of delaying the assailant within the 200-yard zone, so that he would not reach his objective before a great deal more fire could be poured upon him. In essence there were two ways of achieving this – either by turning the attacker around by a Wellingtonian counter-attack in the way we have already analysed, or by the judicious choice of an encumbered battlefield calculated to prolong the firefight.

If the defending troops were insufficiently trained to contemplate a counter-attack, then the obvious way to delay the enemy was to stand behind either natural obstacles or artificial fieldworks. A subtle variant on this theme that was sometimes encountered, whether intentionally or not, was when the enemy was given good cover a short distance in front of the defending line, so he would be tempted to go to ground and open fire. This was what happened to Pack's Portuguese assaulting the Arapil Grande at Salamanca, when they found a convenient breast-high shelf of rock between them and the French. It also happened to Freire's Spanish at Toulouse, when they advanced into a sunken lane but could not be persuaded to advance out of it. In both cases the enemy was able to advance a few leisurely paces, and devastate the attacker with fire from above.[64]

More straightforwardly, however, an attack could be deprived of its impetus if it simply had to climb a steep slope under fire, as Claparède found in the Closewitz ravine at the start of Jena,[65] and Picton's Division found at the start of Orthez.[66] Unimproved slopes usually remained technically passable, however, and in favourable circumstances all but the very steepest could be stormed. Wellington's final assault at Sorauren was up a very steep slope, as were many others in the Pyrenees and more than a few of Napoleon's attacks in the Alps. If he had time, however, a defender could amass additional protection by artificially scarping the natural contours. A flat, near-vertical face some twelve or eighteen feet high might be etched into an otherwise-climbable hillside. The heights around Torres Vedras, for example, still bear the marks of such excavations to this day.[67]

Water obstacles might also be used to delay an attacker, although not even the sea was enough to stop French cavalry from capturing the Dutch fleet across the ice in 1795. Lake Zurich was crossed by Soult's swimmers in 1799, and the Bidassoa estuary, almost half a mile wide, was forded by Wellington's men in 1814. Yet on the other hand a staunchly-defended rivulet could sometimes loom as large as a mighty torrent in the eyes of an uncertain assailant. The tiny Portina brook at Talavera was enough to mark a boundary between the two armies for much of the battle, as was the Goldbach stream on the French right at Austerlitz. Of course it was often possible to increase the value of water obstacles artificially, either by bridge demolitions or by creating inundations, but such measures were more usual in fortress warfare, or in a strategic context, rather than in pitched battles in the open field.

In open battle it was more normal to create obstacles to an attacker's mobility by digging redoubts and other banked fieldworks, surrounded by dry ditches and timber palisades or *chevaux de frise*. The Russian army especially favoured these techniques, and we find it fighting from behind earthworks in many of its defensive battles from Eylau onwards. This did not ultimately stop French infantry from taking Russian positions by storm – and at Borodino even the cavalry managed to do the same – but it did increase the difficulty of such assaults.[68]

Elsewhere the building of redoubts was often associated with hill positions where a long line had to be held by relatively few troops, but where lateral communications were a cause for concern. Torres Vedras and the successive lines of French fieldworks on the Pyrenees both fell into this category, albeit with very different efficacy in each case. In the Portuguese example the fieldworks exerted an entirely successful deterrent effect upon the enemy; but on the Franco-Spanish border they had no appreciable influence upon the battle whatsoever. Time and again Wellington's men were able to storm Soult's apparently impressive Pyrenean hilltop forts, suffering only minor losses.[69] We must conclude that the defenders felt themselves inadequately provided with either supporting echelons of friendly troops, or with personal constitutional fortitude. At all events, it is clear that detached forts were scarcely a very complete answer to the problem of defending an infantry line on an extended battlefront.

Barricaded farms and hamlets were normally more useful bastions of defence, since they tended to be more extensive and could accommodate more troops. On at least one occasion a whole battalion was billetted in a single cottage near Bayonne,[70] and in the same campaign the stalwart defences of Barrouilhet, Arcangues, and St Boës deserve in their way to be no less celebrated than those of Hougoumont and La Haie Sainte at Waterloo.

The classical alternative to a defence based on redoubts, villages or other detached strongpoints was to build a continuous line of fortification, comparable to Hadrian's Wall or the *ne plus ultra* line of Marlborough's era. Similar expedients were tried occasionally during the French Revolutionary Wars, and not without success. The British on the Zijpe line in North Holland in 1799, for example, had exploited the local layout of dykes and canals to establish a flankless perimeter that could be held while they completed their concentration. All it needed was enough men to hold the frontage in sufficient force at every point, and enough time to complete the necessary excavations.

From the Napoleonic period, perhaps the classic case of such a linear fieldwork came in the so-called 'Battle of New Orleans' at the start of 1815, which was in fact a series of four actions spread over two weeks. When examined as a whole, this episode can give us quite a vivid and instructive illustration of many of the points that are most relevant to the tactics of the day. Let us therefore round off this chapter by sailing westwards across the Atlantic from the Iberian Peninsula, to contemplate a battle that – in British eyes at least – has all too often been overshadowed by the events that were soon to take place on the southern outskirts of Brussels.

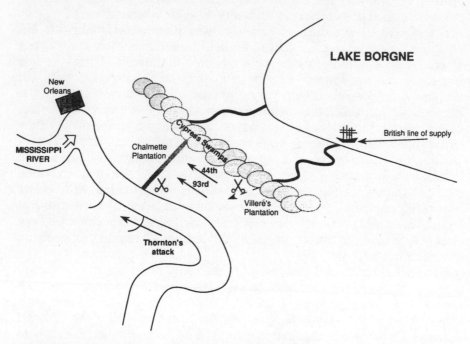

The battles of New Orleans 23 December 1814–18 January 1815.

At the end of 1814 the British had completed their successful attack on Washington and were attempting to advance up the Mississippi through New Orleans.[71] The entrance to the river was blocked by US forts, however, so it was decided to advance around the rear of the city by making a flank move through the waterways further to the East. This 'indirect approach' was initially extremely successful, and achieved an impressive operational surprise. The American flotilla guarding Lake Borgne was destroyed, and General Pakenham was able to land his miniature army undetected, just seven miles from New Orleans itself. The capstone was set on this manoeuvre when Colonel Thornton, the man who had burned the White House, narrowly managed to beat off General Andrew Jackson's shrewd counter-attack at Villeré's plantation on the night of 23–24 December. After this the American forces were thrown into disarray, and for a time there was no coherent defending force between the British and their target.

Ironically enough, 24 December was also the day when the Treaty of Ghent was signed in Belgium between Britain and the USA, bringing the war to a close on terms that were scarcely wonderful for the Americans. Until that time their operations on land had been largely unsuccessful, and a great gloom had descended upon their governing circles.[72] As if by telepathy, however, the British forces outside New Orleans seemed to let go of their grip on reality and military efficiency at the very moment when the treaty was signed. From 24 December onwards, it was the British who were to make all the running in the field of military incompetence.

After their victory at Villeré's plantation, the British entirely failed to make a pursuit, but instead permitted the Americans to steal their herd of draught animals. Jackson was also given time to dig in on a flankless, mile-long position along the north bank of a wet ditch in Chalmette plantation. He used every resource available to improve these positions, but not all of them were felicitous. The improvised cotton bale breastworks unfortunately caught fire from the cannons' muzzle flashes, while an attempted water obstacle equally unfortunately drained off into the Mississippi. Nevertheless, a line of sorts was soon established and roughly manned. At Villeré's plantation Jackson had learned the hard lesson that his shaky forces, half of whom were militia, were unreliable in the offensive. He now spared no effort to patch together a purely static defence. Pakenham had meanwhile arrived with reinforcements, but he greatly overestimated his enemy's strength and decided to wait still longer while artillery was brought forward and a canal opened to provide communications with his boats.

The advance was finally resumed on the 28th, but it was halted as soon as it encountered cannon fire from the Americans' fortified line. The whole British force went to earth in flat but soggy ground interspersed with drains, some 600 yards from Jackson's position. It stayed there, immobile and useless, for seven hours under long-range bombardment. British rocket fire was ineffective in this exchange, and a promising attempt to outflank the enemy was countermanded just in time to snatch defeat from the jaws of victory. Pakenham entirely failed to order any assault, even though at this stage Jackson's defences were still very weak, and he did nothing to boost the morale of his frustrated soldiers.

The scale of casualties in this artillery duel was tiny. With the benefit of fieldworks only nineteen Americans were hit, as against between 60 and 120 of the more exposed British. That makes only between eight and seventeen British casualties per hour, caused by perhaps seventeen American cannon plus a warship on the river – scarcely an impressive argument for the power of Napoleonic weapons! At this range there were probably no casualties caused by American riflemen, who were in any case present only in relatively small numbers. The myth that it was the lynx-eyed, rifle-toting 'Hunters of Kentucky' who beat off the British at New Orleans is based on very slender evidence.

By 1 January 1815 Pakenham had brought up many more guns, and was ready to try again. He successfully achieved surprise, with his infantry making a silent approach through morning mist, without flints in their muskets. However, he did not appear to envisage the operation as an open battle or a *coup de main*, but rather as the opening of a formal siege. He used his advantage only to set up batteries at a prudent range from the enemy – not to charge forward over the puny entrenchments. This was a major lost opportunity, which depressed the British troops still further at the same time as it gratuitously handed a boost in morale to Jackson's ragged and disparate band.

The exact time taken in cannonading on 1 January is open to dispute, but the British troops lay out in the open all day while the rival artilleries attempted to silence each other at ranges between 600 and 800 yards. Overall, the British lost 67 casualties and had to abandon five guns, inflicting a mere 22 losses on the enemy. On the inland flank there was a close action when Coffee's Tennessee riflemen beat off a small infantry attack, but this can scarcely be called a triumph of the rifle's special properties. Coffee's cleared field of fire was only 30 or 40 yards wide, which falls well within the range of smoothbore muskets.

Pakenham's first two approaches to the US line had both been given limited objectives that stopped short of storming the entrenchments, but the Americans were well and truly alerted as a result. They raised the level of their mud parapets – reaching nine feet in places – and they received sizeable reinforcements. The British, by contrast, were becoming progressively more demoralised by their officers' bungling and by an epidemic of dysentery in camp. On 8 January Pakenham nevertheless determined to make a third approach, in two columns linked by riflemen, covered by morning mist and – at long last – with the aim of making an escalade at close quarters.

The assault was comprehensively botched. It was not properly reconnoitred, and its start was damagingly delayed due to a vain attempt at co-ordination with a secondary attack to clear the west bank.[73] This meant that visibility had begun to improve when the forward movement commenced, and the American artillery was able to open at some 500 yards' range. Then it was found that the 44th regiment, leading Gibbs' right-hand column, had failed to bring either its fascines or its scaling ladders. It fell into disorder close to the enemy's ditch, and its supporting regiment was left to make an unexpected passage of lines to reach the front. Simultaneously the 93rd regiment, leading Keane's left-hand column, was inexplicably ordered to halt without firing, within 100 yards of the enemy's breastworks.[74]

The British had been told not to fire because that might have given away their position during the approach march, and then might have stopped their impetus once the assault had been launched. When they had halted in front of the enemy, however, there was clearly no longer any justification for this policy, and the soldiers were simply shot down like rabbits without doing anything in return. If more of them had fired back, the Americans might have suffered more than their actual loss of at most 52 killed and wounded.

On the American side there was apparently also a desire to hold back musketry until very close range, and certainly less than the range to be expected with rifles. Jackson is reported to have ordered fire 'at fair buck range', aiming 'for buckles on the crossbelts'.[75] In the event, however, the Americans seem to have opened somewhat earlier than intended, with the consensus of estimates setting the range somewhere around 200 yards.[76] The colonel of the 44th appears to have been sniped by a rifleman at 300 yards,[77] but apart from this, most of the rifle fire seems to have occurred at point-blank range ('fifty feet') *after* a small British 'forlorn hope' had managed to enter the most westerly redoubt. This fire came from Beale's company of 35 New Orleans gentlemen-sharpshooters, who seem to have been almost the only rifle-armed Americans to be involved in the main centres of combat.[78]

As for the duration of the firing, it is variously set between five minutes and half an hour, probably indicating a major crescendo at first, trailing off into spasmodic shots as the British withdrew. It did not all take place at the opening range, but closed to less than 100 yards on both of the two main fronts – and individuals advanced to point-blank range at a number of places. British riflemen from the 95th regiment had initially been posted as skirmishers between 100 and 150 yards of the enemy before the main action began, and advanced to give close covering fire from the very foot of the fieldworks – but in the absence of a determined effort by the main body they do not appear to have achieved a great deal. Even if the entire US loss is attributed to the five British rifle companies firing only two shots per man – unrealistically assuming only a five-minute engagement – then we still find that it took around twenty rounds to hit each American. This certainly seems to fall short of the inflated expectations which some modern writers would have us entertain from the rifle.[79]

The casualties were enormous on the British side, certainly rising above 2,000, and possibly above 3,000, out of the 5–6,000 directly engaged. Most of the important officers were incapacitated, including Pakenham himself. He had quite correctly moved into the front line to urge the hesitant troops to resume their assault, and when he found they could (or would) make no progress with that, he called for the reserve brigade to advance and carry the position in a second impetus. It was at just this moment that he was killed, an event which Lambert, his second in command, took as decisive for the battle as a whole. He immediately cancelled the movement of the reserve, just when it might have been most effective.

New Orleans has often been taken as proof that a ditch and embankment topped by riflemen, provided its flanks were secure, was an unbeatable formula for defence during the Napoleonic period. The evidence, however, scarcely supports this interpretation. We have seen how the rifle was a negligible factor in the American victory, since most of the fire came from either cannon or smoothbore muskets; nor did their own riflemen appear to have been much help to the British. Essentially, the same result would have ensued if there had not been a single rifle on the battlefield.

As for the fortifications, there are plenty of signs that they could have been scaled, if only the attacking troops had pushed forward and actually tried to do it, even without the help of fascines or ladders. The 93rd could certainly have achieved more impressive results than it did, if only it had not been prevented from doing anything by its officers. By contrast, the 'forlorn hope' at the western end of the line actually took a redoubt almost before it could fire back; while elsewhere a number of individuals surmounted the works and were captured on the American side of them. Quartermaster Surtees' riflemen reported they could have passed the ditch 'with ease', and it was his own opinion that if the 44th had pressed forward rapidly.

. . . their loss would not have been half so great; for the enemy's troops in front of the right column were evidently intimidated, and ceased firing for some seconds as the column approached; and there is little doubt, had they pushed on to the ditch with celerity, the Americans would have abandoned their line . . [80]

Captain Simpson of the 43rd regiment believed that

. . . protected by their [*the 44th and Rifles*'] fire, it was quite possible to have accomplished the passage of the ditch without the assistance of either scaling ladder or fascine . . . This observation is made in consequence of the whole of the defences having been passed by me, partly as conqueror, and partly as a wounded prisoner.[81]

48

Perhaps the key to these failures is that the spearheads of the assault columns were composed of relatively inexperienced units. Contrary to the claims of Boston propaganda, neither the 44th nor the 93rd had taken part in the Peninsular War. The 93rd, in particular, had enjoyed a quiet, sun-drenched decade since 1805 in Cape Colony. Apart from the 95th rifles, however, Pakenham deliberately took the decision to keep his Peninsular veterans either in second line or in reserve. They saw relatively little of the action, and were therefore effectively wasted. Surtees is doubtless right to claim that they could have rolled over the American line with little loss, if only they had been placed in the van;[82] but they were not, and the moment was lost.

The true lesson of New Orleans is that inexperienced troops are vulnerable if asked to manoeuvre or make difficult assaults. The Americans had found this on 23 December, and the British on 8 January. Yet if they were properly supported by good artillery, fieldworks and leadership, the same troops were capable of laying down a heavy blanket of musketry at close range – especially if the enemy obliged by halting in the open. What is not proved by New Orleans, however, is that firepower on its own could stop an assault by experienced and strongly motivated soldiers, even against artillery and fieldworks. Apparently it took a very special sort of high command bungling to do that.

3

1815–1915: The Alleged Novelty of the 'Empty Battlefield' in World War I

Approximately half of the casualties suffered by the French Army in the First World War came during the first fifteen months of the conflict. It was during these months that it became obvious that the war, against all expectations, was to be a long and expensive affair. It would not be finished in a couple of mobile campaigns, as had been hoped. Instead, it was to be a contest of attrition in which the manpower, industry and morale of each nation would be tested to the limit and sometimes ultimately broken.

This realisation came as a profound shock to almost all participants, and it soon generated bitter feelings of betrayal and recrimination. The legacy of those feelings has been faithfully handed down to every generation until the present day, since it seems that a miscalculation made in 1914 has led to nothing less than the spoiling and bloodying of the entire twentieth century.

The miscalculation which made the war indecisive was purely military in nature, and indeed largely a matter of minor tactics. It arose from the erroneous belief that good infantry could, as often as not, capture enemy defensive positions and hold them against a counter attack. If this feat had proved possible, then it would also have been practical for entire offensives to make steady progress leading to decisive break-outs behind the enemy's lines. In the event, however, it did not turn out to be possible. The successful capture of an enemy position became a very rare event indeed.

It seems that the generals of 1914 failed to appreciate the significance of enhanced firepower which a variety of new weapons had placed in the hands of their soldiers. They did not fully realise the havoc which could be caused by the combination of mud, barbed wire, magazine rifles, machine guns and quick-firing artillery with high explosive shell. There was apparently a misplaced confidence that although all this might well increase the casualties suffered by attacking troops, such losses would rarely be enough to affect the final outcome. After receiving a few successful attacks, it was felt, a defender would realise the hopelessness of his posture and retire. He would then suffer the much higher losses incurred by an army in flight. The outlay in lives which the attacker had made would be more than recouped, provided that his offensive morale had been initially high.

We must therefore ask what caused this apparent myopia on the part of so many generals who were, after all, the chosen leaders of a reasonably technical profession. By 1914 they were supported by a wealth of deep and perceptive staff studies, and far more good specialist advice than had ever been available in the days of Napoleon. Science and industry were bombarding the armies of Europe with their products, which the armies themselves were not slow to accept. It is therefore all the more astonishing that such a unanimous miscalculation should have been made. Why was it made?

The normal answer to this question is that the generals were actually stupid and pig-headed. They had political interests which clouded their analyses, or social backgrounds which prevented them from under-standing the revolution being wrought in the laboratory or steel-mill.[1] One could not expect a Prussian Junker or an Anglo-Irish landlord to understand all the obscure technological developments which seemed to succeed each other so rapidly. Armaments were changing with such unprecedented and vertiginous speed that it was hardly surprising if the officer caste was left behind.

There is undoubtedly a great deal of truth in these charges; but at the end of the day they can give us no more than half of the total picture. As so often in tactical studies, sociological or political explanations cannot really get to the root of the question. There is also a story which must be told in terms of purely military logic; an evolution of theories and experiments which finally led decision-makers to believe that they had no option but to adopt the course they did.

It is particularly misleading to think of the tactical changes of 1914 as a 'revolution', or something which happened suddenly. The generals of the First World War were not confronted by a totally new or unforeseen set of circumstances, but had been given astonishingly long advance warning. In both theory and practice the debate about firepower had been going on ever since Waterloo, and possibly even earlier. Anyone concerned with formulating tactical policy had found it thrust unavoidably under his nose.

What happened in 1914 was that armies finally put into practice the conclusions they had reached after many years, and in fact many decades, of deep thought. They were perfectly well aware of the problems posed by new weapons; but believed they had found the scientifically correct solutions. They had examined all the options and chosen the least costly. It is for this reason, and not because they were 'out of their depth', that generals clung so stubbornly to the methods they had chosen. All other methods seemed to offer far more dubious results, if not total disaster. In a sense we might even suggest that the generals of the First World War felt they were using advanced professional expertise to minimise the losses. They were doing all the things they had trained to do, with good prior knowledge of what the problems would be.

These problems are best summed up in the expression 'the empty battlefield'. With improving firepower infantry would no longer be able to show itself within range of the enemy in heavy formations, since they would make too good a target. Instead, the troops would break down into loose chains or skirmish screens, and seek to use the terrain for cover. All an observer would see would be the occasional head, peering out and looking for a target. Individual marksmanship would become much more important in a battle of this type, and the range of engagement would increase. A true musketry 'firefight'[2] would take place, in which the infantry would try to kill its enemies at a distance. Gone are the heady massed scrimmages so beloved of Victorian battle painters: the battlefield starts to look more like an empty field than a drill parade.

It has often been assumed that the 'empty battlefield' first appeared in the First World War, or at least in the later nineteenth century. The introduction of smokeless powder, in particular, has been considered a decisive event, since riflemen no longer needed to give away their positions when they fired. This innovation coincided with the appearance of the high explosive shell, the perfected magazine rifle and the Maxim gun in the late 1880s, and so it has been acclaimed as the turning point by those who would look for a 'revolutionary moment' in the evolution of tactics.

It is true enough that all these developments were important; but in fact they mark no more than one evolution among many in a continuous process which had started very much earlier. The idea of the empty battlefield was already well known long before the late nineteenth century, and the problems which it posed had already been extensively analysed by tacticians.

Let us consider, once again, the Battle of Waterloo. We have already seen the methods of action adopted by massed bodies of prime infantry when they came to close range; they relied upon the bayonet rather than upon musketry. For them the objective was to scare the enemy into panic flight rather than to kill him. If they did get into a close-range firefight like Maitland's, then it would be a brief but bloody exchange between two closely-packed masses. On other parts of the field, however, there was an altogether different type of combat being fought, in which the infantry was less ready to come to close contact or to use massed formations, and more anxious to use sustained aimed fire and the protection of the ground.

Captain Leach of the 1/95 Rifles gives us a clear description of a part of the field which was to all intents and purposes 'empty'. He reports that:

From the time that La Haye Sainte fell into the hands of the French until the moment of the General Advance of our Army, the mode of attack and defence was remarkable for its *sameness*. But I speak merely of what took place immediately about *our* part of the position.

It consisted of one uninterrupted fire of musketry (the distance between the hostile lines I imagine to have been rather more than one hundred yards) between Kempt's and some of Lambert's regiments posted along the thorn hedge, and the French infantry lining the knoll and the crest of the hill near it. Several times the French officers made desperate attempts to induce their men to charge Kempt's line, and I saw more than once parties of the French in our front spring up from their *kneeling position* and advance some yards towards the thorn hedge, headed by their officers with vehement gestures, but our fire was so very hot and deadly that they almost instantly ran back behind the crest of the hill, always leaving a great many killed or disabled behind them.[3]

He added that he believed this to have been 'the closest and most protracted musketry contest almost ever witnessed'; and indeed it appears to have continued for some hours on end. Not the least significant aspect of this fight was the great length of time during which it continued.

It is clear from Leach's report that both sides in this duel were using cover – the hedge on the British side, and the hill crest on the French. It is also likely that the troops on each side were in a fairly loose formation, since the British rifle regiments were always accustomed to fighting in this way, while the French were firing from a kneeling position. It is impossible to conduct a kneeling fire in any very dense formation, and it would appear that in this case the French must have had a depth of no more than one rank. When they did make forward rushes it was only in small groups, so they cannot have been arrayed in anything like the solid masses used in the earlier attack by D'Erlon's divisions or the later attacks by the Imperial Guard. The French force was probably composed of tired and defeated troops who had regrouped from D'Erlon's first abortive assault, who wanted to keep up the fight, but who lacked the cohesion or unity of a fresh massed formation.

What Leach gives us is nothing less than an account of an empty battlefield in the Napoleonic period. The range was perhaps rather short by late nineteenth-century standards, because the smooth-bore musket could not reach much further than one or two hundred yards. By comparison with encounters between fresh masses, however, the range was great enough to allow the combat to drag on for a long time.

Sufficient casualties were inflicted at a distance to deter attackers from pressing home; yet they were not enough to produce a decisive result. This was a genuine firefight in which both sides tried to kill the enemy rather than merely to scare him off: the 'battle of killing' had already begun in this part of the field. We have the testimony of another rifleman, Captain Kincaid, that Waterloo seemed extraordinary because it had the aura of a 'battle in which everybody was killed'.[4] Compared to many other Napoleonic battles he felt that there was an excessive proportion of dead to fugitives at the end.

We must ask whether Leach's combat at Waterloo was really as exceptional as he seemed to think. Was it no more than a unique oddity, or does it reflect a type of activity which had been going on unnoticed in many other Napoleonic fights? To what extent was the battle of killing an important part of all the warfare of the age?

To find an answer to these questions we must look at the 'Eastern Front', where the French fought Prussians, Russians and Austrians, rather than at the Peninsular campaigns where they faced the British. In the Peninsula there was certainly a great deal of skirmishing, and some of it was highly professional on both sides. The British rifle regiments, in particular, were celebrated for their skill in maintaining flexible screens ahead of the main fighting line, and in picking off French officers at long range. In the Peninsular battles, however, there seem to have been relatively few sustained firefights compared with the head-on collisions of formed bodies. The skirmishers were an auxiliary arm, designed merely to pester and confuse the enemy as he marched into the contest at close range. There were some notable exceptions to this rule, as at Arcangues church in 1813, or in Craufurd's defence of the Coa bridge in 1810; but by and large the battles seem to have been decided in fields which were less than 'empty'.

In Eastern Europe after about 1808 on the other hand, conditions were rather different. On average the opposing soldiers tended to be less hardened to campaigning than those of the Peninsula, and more reliant upon their artillery and skirmishers. Armies were bigger, contained a higher proportion of raw levies and were capable of less precision in manoeuvres. In these circumstances it was often more difficult to bring massed bodies into close combat, so an alternative method of defeating the enemy had to be found. That method turned out to be fire combat.

Clausewitz, one of the most eminent veterans of the 'Eastern Front', suggested that there was a clear distinction to be drawn between the fire combat and close combat phases of a battle. He said that the first was designed primarily for the physical destruction of the enemy, whereas the second was for the destruction of his morale. The fire combat should therefore be seen as an effective and essential preparation for the decisive bayonet charge, and should preferably be conducted by different troops from those who would be sent into the charge itself. It should not be seen only as a perfunctory auxiliary, but had to be a serious and protracted business, since significant physical destruction of an enemy's force could not be achieved quickly with the weapons of the day. Of especial interest to us here, he added that this firing or destructive phase of the battle would take a length of time proportional to the size of the forces engaged; thus a big army would require a much longer firefight than a small one. In the conditions of Eastern Europe, in other words, the preliminary exchange of fire would be conceived on a much greater scale than in the relatively small battles of the Peninsula

Clausewitz drew a vivid picture of what happened to the soldiers involved in a firefight of this type:

> After a fire combat of several hours' duration, in which a body of troops has suffered severe loss (for instance, a quarter or one-third of its numbers) the debris may, for the time, be looked upon as a heap of burnt-out cinders, for – (a) The men are physically exhausted; (b) they have spent their ammunition; (c) their arms want cleaning; (d) many have left the field with the wounded, although not themselves wounded; (e) the rest think they have done their part for the day, and if once they get beyond the sphere of danger do not willingly return to it; (f) the feeling of courage with which they started has had the edge taken off, the longing for the fight is satisfied; (g) the original organisation and formation are partly destroyed, or thrown into disorder.[5]

He said that troops in this state would be in a condition neither to deliver nor to receive a bayonet charge. For them the battle would have been entirely a matter of fire action with an attritional rather than a decisive aim. He went on to suggest that in this phase of the combat a good general would try to put as few soldiers as possible into the firing line, since a relative advantage might be gained by exposing only a few dispersed musketeers against a larger but more concentrated body. Both sides might lose the same number of casualties in such a fight; but the larger force would have more men exhausted by the end. It would have burnt up its potential reserves, while its opponent might still have fresh supports waiting for the decisive moment in a covered position further back.

Clausewitz does not claim that all Napoleonic fighting was conducted by fire, let alone decided by it; but he does imply that the firing phase could drag on for much longer and involve more troops than was normal in the battles between British and French in the Peninsula. He stresses the need to maintain and feed a skirmish line during extended periods of deadlock between bayonet onsets. As if in confirmation of his view, we do indeed find that in some central European battles entire divisions of infantry were broken down to skirmish, while in all armies after about 1808 there was a much increased provision of troops allocated to skirmishing duty. The Prussians in particular took this matter seriously, and allocated over a third of all their infantry as skirmishers.[6] Some of the Eastern European battles lasted for two or even four days on end, of which only a small proportion can have been taken up by bayonet attacks. Much of the interval must have been given over to the firefight.

Skirmishing infantry was not the most destructive source of firepower on the Napoleonic battlefield, however, because artillery was starting to be used regularly in very dense formations. Sometimes batteries of up to a hundred guns would be concentrated, which could sweep the ground in front of them of all opposition.[7] Many of the biggest battles were to a very considerable extent 'artillery battles'; and it would certainly be misleading to think of this particular feature as an original invention of the First World War. For contemporary commentators, however, even the appearance of massed artillery seemed to be overshadowed by the massed use of skirmishers. For a variety of reasons it was this, and not the artillery, which most seized the imaginations of military theorists.

There cannot have been anything particularly surprising in the fact that artillery firepower was gradually increasing. The artillery was a self-consciously technical arm, and in most countries it had committees, test ranges and factories dedicated to the task of improving performance. With the infantry, on the other hand, there was none of this scientific paraphernalia. The infantry was generally thought to require the least technical training of any arm, and was not felt to be susceptible to radical developments.[8] It was judged to be good or bad purely on the basis of its resilience on campaign and its mastery of the close-order drill book. Beyond that there seemed little to say.

With the arrival of massed skirmishers, however, the intriguing possibility of an 'infantry science' seemed to be opened. A specialist skirmisher needed a number of qualities which the traditional line infantryman apparently had not. The skirmisher had to be able to aim his piece at the target, so the whole science of ballistics and marksmanship became relevant to him. He had to be able to move quickly around the battlefield, so the nascent science of gymnastics was important. Of particular interest, he had to be able to make his own decisions. He was no longer under the close control of a solid rank of NCO's and officers: he was on his own. This meant that ways had to be found for developing the intelligence and understanding of every single skirmisher. Education, moral instruction and even political consciousness were invoked as necessary aspects of a skirmisher's training. Such soldiers would not be mere 'automata' like the members of Frederick the Great's mechanically-drilled line. Instead, they would be what the French called 'baionettes intelligentes'. They had to be able to think for themselves.

In Britain Sir John Moore had already taken some important steps along this path with his light infantry training programme at Shorncliffe after the Egyptian expedition. He stressed not only the importance of firepower and a dispersed formation, but also the need for a more humane discipline which would give free rein to the personal qualities of each soldier. He treated his men as individuals, and encouraged the development of their personal initiative.[9]

All these arguments suggested a complete revolution in the role of infantry, and indeed in Prussia the skirmishing lobby even came to be associated with the idea of revolution in the political sense.[10] Far from being the most reactionary and authoritarian arm, the infantry now seemed to require an influx of progressive, liberal and scientific ideas. It was perceived as a veritable spearhead of development in the military field, and junior officers of ambition were delighted to find that it could offer them something more than an endless round of drill.

In technical terms, also, this period saw a number of important developments. The British rifle regiments with their Baker rifles, as well as the rifle guilds of the Tyrol, proved that accurate marksmanship in battle was possible at greater ranges than had been practicable with smooth-bore muskets. After a steady series of improvements from 1815 onwards, the Minié bullet and the percussion rifled musket were brought together in the 1850s to make a cheap but very effective weapon for mass use.

There was also talk of a military breech-loader. Such a weapon had actually been used by Colonel Ferguson in the American Revolutionary War, although it had finally been judged too fragile and specialised for general adoption.[11] Between 1814 and 1822 the French conducted a series of tests with another model, which was finally rejected by only a narrow margin – again on the grounds of fragility.[12] In 1841 the Prussian army actually did start to deploy breech-loaders, in the shape of what Mr. Punch laughingly called the 'death defying Needle-gun'. Under campaign conditions the Dreyse Needle-rifle, like its predecessors, proved to be less than totally soldier-proof; but in 1864 and 1866 it did give certain tactical advantages to its users. The decisive defeat of the Austrians at Sadowa, in particular, was thought by some observers to be due largely to the prone firing posture which the new breech-loaders made possible.

As if all these improvements were not enough, there was also the realisation that a practical machine-gun was just around the corner. In his important book *De l'Armée Selon la Charte* (1829), General Morand discussed the uses to which existing steam-driven machine-guns (e.g. the Perkins model) could be put. He admitted that the encumbrance of their boilers would render them unsuitable for mobile warfare; but he suggested that their smokeless discharges would be ideal for defending towns. An enemy could be lured forward down an apparently deserted line of approach, and then ambushed with automatic fire from concealed positions covered by extensive obstacles:

Bombardment (of the town) need not be feared, since the defence would be conducted at a distance, hidden in the ground, in casemates or dug-outs. Because these automatic weapons would produce neither noise nor explosion, would operate with hardly anyone to man them, and would be concealed by tufts of grass or bushes, the enemy would be unable, in his emotion, to identify the source of their invisible shots.[13]

Having achieved the initial killing surprise, the defender could then send infantry forward from their trenches to finish off the business with a counter-attack.

The above description comes from the pen of a leading tactical writer of the immediately post-Napoleonic period. In essence he is advocating a genuinely empty battlefield, in which machine guns and camouflaged trenches replace the massive masonry ramparts of existing fortress architecture. Strong echoes of the Western Front resound throughout this passage, and we are forced to admit that more famous prophecies of the First World War such as H. G. Wells' *The Land Ironclads* (1903) actually came rather late into the field.

If we now return to the way in which the Napoleonic skirmish line was viewed by contemporaries, we can start to appreciate the full scope of the revolution which it promised. The front line of infantry would in future be dispersed, individually self-reliant and able to kill the enemy at a much greater range than hitherto. It would even be able to defeat artillery by its fire, in circumstances where the gunners had previously enjoyed the upper hand. More than one writer called the new generation of rifles 'hand artillery', insofar as a battalion armed with them could lay down a greater number of bullets in a given time than could a battery of artillery at the same ranges.[14]

If the battlefield supremacy of artillery was being called into question by improved infantry weapons, so also was the role of cavalry. An important element in the 'skirmish lobby's' arguments turned on the rapidity of infantry movement around the battlefield. Whereas Napoleonic infantry had walked everywhere until the final charge, the new infantry would run. This would give it a mobility almost equal to that of the cavalry, but with considerably greater firepower and at a dramatically smaller cost to the taxpayer. It would also be able to shoot down attacking cavalry at long range and could safely rely upon small rallying squares for close-range protection, rather than the ponderous battalion or brigade squares of the past. For the champions of light infantry the cavalry was already an outdated weapon as early as 1840, and by the time of the American Civil War some infantry units were starting to be called 'foot cavalry'. In the Russo-Japanese War of 1904–5 Sir Ian Hamilton noticed that Japanese infantry seemed able to attack at no less than four times the speed of their European counterparts:

This combination of infantry tactics with cavalry speed of movement; which might, as I have said, demoralise second rate troops, might also considerably disconcert even the very best of marksmen.[15]

The new infantry was supposed to fight a 'battle of killing' dispersed and at long range, which they might then follow up with a very rapid charge across no-man's-land. They would move so fast that they would be exposed to relatively few of the enemy's shots, and would demoralise him by their onset to the extent that his aim would be spoiled. Colonel Le Louterel summed up this effect when he wrote in 1848 that '. . . in front of the enemy, the emotion and internal misgivings which are inseparable from a real combat will always prevent a soldier from shooting accurately'.[16] Whereas fire with modern rifles could be destructive at long range, there was normally little to fear from a firing line which was being attacked violently at close range.

All these ideas were brought together during the late 1830s in the French Army, when the first units of 'Chasseurs à Pied' (originally 'Chasseurs de Vincennes' and then 'Chasseurs d'Orleans') were formed. This was to be 'an infantry which is truly light'.[17] It was to be capable of sustained movement at the 'gymnastic pace' or jog, yet at the same time it was to be trained in accurate shooting with the rifle. The Chasseurs would also be encouraged to develop individual initiative and education. Their champions saw them as a veritable blueprint for the army of the future, and predicted that they would soon put all other types of soldier out of business.

By 1853 there were no less than twenty battalions of Chasseurs, and they regarded themselves as a separate arm of the service. They attracted many of the more promising officers of the infantry, an exceptionally high proportion of whom later rose to become generals. It was even true that certain elements of Chasseur training came to be adopted by the army as a whole. Education for personal improvement, gymnastic exercises and target practice all came to be generally accepted; and around this period the rifle was itself starting to be distributed to the troops of the line. Foreign observers were mightily impressed, and indeed no less an authority than Friedrich Engels waxed almost lyrical about the individual freedom, firepower and mobility which this new and democratic weapon seemed to embody.[18]

In France as elsewhere, however, there was also a reaction against the Chasseur movement. Precisely because it concentrated a large number of effervescent and innovative young officers in one place, it was automatically distrusted by many of their elders. Because it represented an attack on the artillery and the cavalry, equally, it could not fail to make many enemies. Of particular importance, perhaps, were some of the moral and political assumptions which lay behind the light infantry movement. No longer was the infantryman to be fully under the control of his officers.

He was to be master of his own fire – and indeed the Chasseurs had originally dispensed with the word of command to 'fire' altogether.

For some officers this devolution of decision-making was pure heresy. In the Infantry committee of the French Ministry of War in 1844, for example, we hear that:

Several members objected to the proposition that soldiers should be the absolute masters of their muskets. The soldier was considered to be no more than a rack on which the musket could be rested.[19]

Perhaps rather less extreme but more widespread was the view of Captain Boyer, who wrote in 1833 in favour of close formations and tight 'surveillance' of the men by their officers and NCO's:

By concentrating your defences into a smaller space . . . everyone is forced to do his duty. In this order of battle the officers will acquire a great ascendency over the spirits of the men, whose morale is easy to excite on the battlefield, and especially when leaders close at hand give them an example of devotion and contempt for danger.[20]

It was especially the morale and obedience of the soldiers which open order was thought to undermine. Skirmishing with long-range rifles was considered to be indecisive – as indeed it had been in Napoleonic times – and therefore to be ultimately demoralising. 'One would think that their [i.e. rifles'] use would be favourable neither to the courage in action, nor to the intelligence, which have in every age ensured the superiority of the French soldier.'[21]

There was a great feeling in the French Army that although long range fire might achieve certain indisputably concrete results, it would do so only at a high psychological cost. There was absolutely no point in killing the enemy if at the same time you also destroyed your own men's will to advance. Simple humanity seemed to dictate a less indecisive form of action, as did military expediency.

An official answer to the 'battle of killing' of the Chasseur school gradually emerged in the French Army, which stressed heavy firepower at short rather than long range, combined with a vigorous and decisive bayonet charge. This formula had similarities with the British Peninsular technique, although in the French Army the battalions were generally to be formed in columns rather than in lines. The important thing was seen to be keeping the troops under the control of their officers in close formation for as long as possible. Only if this were done could they function with precision and decisive effect.

We have already seen that when he discussed British tactics Jomini advocated a combination of close range fire with a bayonet charge. Of even greater influence, however, were the writings of Colonel Bugeaud, who later became a Marshal of France. Bugeaud specifically laid down the need to avoid indecisive 'tiraillements' between firing lines which

60

would only exhaust each other to no purpose. Instead the troops should advance in close order with only a few skirmishers preparing the way. A single volley with double-loaded muskets followed by a controlled forward rush, would be enough to disperse the enemy.[22]

For Bugeaud the true science of the infantry lay not in the target practice of the Chasseurs, but in techniques for fostering high morale and preventing the spread of disorder. There was common ground between the two camps in that they both recognised the need to educate the soldier and develop his skill in rapid marching; but whereas the Chasseurs called for individual initiative, Bugeaud was a firm authoritarian. For him the cohesion of a unit was entirely the responsibility of its officers, and once that principle was fully accepted everything else would be easy. His Napoleonic experience led him to believe that most infantry degenerated too quickly into a formless rabble. A well led French battalion which could maintain its formation, therefore, would be more than a match for all comers:

A troop thus led will always be brave and rarely defeated, because it will rarely encounter enemies with equal elevation of spirit and equal principles of combat.[23]

Bugeaud's work today appears to be markedly Napoleonic and backward-looking in tone, yet at the time it was written it seemed inspired by daring modernity. It provided an answer to the technical arguments of the light infantry lobby, by appealing less to the past than to brand-new 'sciences' of the Romantic Age. Bugeaud ignored the formal geometrical calculations which had for long been central in tactical writing. Instead, he analysed the 'moral side of combat', and seemed to be looking deeper into the psychology of soldiers in battle than any previous writer. His talk of deployed lines and a single volley at close range also sounded highly un-Napoleonic to officers who were used to regimental columns and protracted firing between skirmishers: Bugeaud's system was therefore accepted as a step forward which combined old and new elements into a fresh and constructive vision.

Our attitude towards early nineteenth-century armies has perhaps been clouded by the belief that they knew nothing of science. We can too easily assume that before the age of mechanisation all generals were opponents of anything which smacked of technology, industry or (worst of all) towns. This picture is not quite correct, however, since although there was certainly a strong ideal of rural nobility among these men, there was also a genuine realisation that they were living in a scientific age. In France the cosmology of Laplace was enjoying an astonishing vogue which even brought it into the curricula of regimental primary schools. The high pace of technical change has already been mentioned in the field of armaments; but it also spilled over into many other types of military equipment such as maps, clothing and transport. All these things were changing rapidly and the achievements of science were on everyone's lips. Even the most reactionary officer from the remotest Vendean farmyard must surely have been aware that times were moving in a technical direction.

Soldiers even started to appreciate that science could bring promotion. In some cases this was true to a limited extent at a personal level, as for the infantrymen Delvigne and Minié, who both invented improved rifles. It was much more true, however, at the institutional level. In the eighteenth century the so-called 'technical arms' of the artillery and engineers had gained an influence in the army which far exceeded their numerical strength. Not only on the battlefield but also in ministerial committees had these branches been able to multiply numbers by technology. Their privileged position was much envied by other arms with less science at their disposal.

After the Napoleonic campaigns, however, warfare seemed to have changed. It was felt that in the new age there would be a greater opportunity for other arms to rival the artillery and engineers in their scientific achievements. Professional staff officers, especially, pointed out that the 'science' of strategy had now come of age, giving the staff a natural authority over other branches.[24] The infantry also tried to develop a variety of 'sciences' of its own, as we have seen; while the cavalry equally plunged into an ambitious programme for applying the new 'sciences' of horse breeding and horse care. When Field Marshal the Earl Haig made his famous statement after the First World War to the effect that the well bred horse would always have a place in battle, he was doubtless echoing the cavalry's belief that it had kept up with the march of technology just as much as any other arm. The 'well bred horse' of the early twentieth century was indeed a very different animal from the nags deployed by Murat and Lassalle. Applied science had made it so, and the cavalry therefore believed itself to be numbered among the 'technical arms'.

Far from being unaware of science, early nineteenth-century armies did everything they could to invent new sciences. Their problem was that there was too little genuine innovation to go round. Many branches of the service had to make do with pseudo-sciences such as gymnastics or horse-care. This did not diminish the ardour with which such activities were pursued, but rather increased it. Ever greater efforts were made by the sponsors of new techniques to demonstrate their scientific status. The first training course for French regimental gymnastic instructors, for example, lasted three years and included a number of theoretical subjects at university level. It was perhaps hardly surprising that some of the pupils left the army in disgust, while the remainder committed acts of indiscipline against their instructor.[25]

It is in this context that Bugeaud's emphasis on high morale should be seen. He was effectively offering the line infantry a new self-image which would allow it to claim equal status with the Chasseurs and even with the more traditional technical arms. He was implying that morale-building was a science which could bear comparison with any other. However seriously this may have been received, there was undeniably a great sense of innovation and change about the infantry throughout this period. It was conscious of doing more for each individual soldier than had ever been done before. As a long-service conscript he was seen as the best possible compromise between the citizen and the professional. He embodied all that was best in the nation, and deserved far more than a deadening round of drill. His enthusiasms were to be aroused and maintained by better barrack conditions, lighter discipline, regimental education courses, gymnastics, fencing, target practice (a notable innovation, this!), and even singing and dancing. The 'spirit of Sir John Moore' was powerfully at work in France at this time, albeit without his central theme of fighting by long range fire.

In practice the Chasseurs à Pied found that their skills were never fully understood or employed on the battlefield. Commanders tended to regard them as élite assault troops rather than as skirmishers. They were thrown into a van of attacks and suffered disproportionately heavy casualties. On the other hand they did at least have the satisfaction of seeing some of their morale assumptions gradually absorbed into the army. Throughout the remainder of the century there was a strong emphasis on the need to educate the soldier, as well as a steadily growing acceptance of open-order formations. In the short term, however, it was Bugeaud's slightly different emphasis which predominated.

In 1854 Bugeaud's instructions were issued almost verbatim by Marshal St. Arnaud to the troops about to fight in the Crimea. On the battlefield they seemed to have worked fairly well, although in confused fights such as Inkerman it was already being realised that soldiers might use a lot more individual initiative than had been foreseen. The Crimea was perhaps not a fair test, however, since the French and British enjoyed a marked superiority in both armament and tactical flexibility (Balaklava notwithstanding). The Russians tended to manoeuvre in excessively deep columns, and did not possess the rifle-musket.

A rather more serious test came in 1859, when for the first time the French encountered an enemy who was himself fully equipped with modern weapons. In this case it was the Austrians in North Italy, with their Lorenz rifles. French tacticians believed that this challenge would require a somewhat altered response, and in some respects they moved away from Bugeaud's prescriptions. While retaining his emphasis upon morale and leadership, they felt that in this war there should be a greater stress upon rapid forward movement through the zone beaten by the enemy's fire. Battalion columns covered by skirmishers were therefore to be preferred to Bugeaud's deployed lines, and the soldiers making the charge itself would no longer be expected to stop to give a volley as they came. There was an increased fear that troops which stopped to fire would become bogged down in an indecisive exchange of shots. Everything possible had to be done to come to close quarters.

In his tactical instructions of 1859 Napoleon III made a classic statement of what he wanted his men to believe:

The new weapons are dangerous only at long range; they do not prevent the bayonet from being, as in the past, the terrible arm of the French infantry.[26]

The astonishing thing about this apparently optimistic formula was that it worked beyond the wildest dreams of its architects.

The 1859 Italian campaign was possibly the most important of all the steps which led to the tactics of the First World War. In this campaign the leading military nation of Europe devised and executed a successful technique for overcoming an enemy armed with rifled muskets. It was a technique which rested not upon firepower but upon the bayonet. High morale and rapid movement in small columns were found to make an effective answer to improved firearms. The French took this lesson to heart, and felt that it finally resolved the debate about firepower which had been raging since 1815. It was therefore small wonder that they showed little interest in the tactics of the American Civil War a few years later.[27] Why should they, since the Americans were only tackling a problem which the French had already solved?

The central fact of the Italian campaign was that charges with the bayonet were able to carry enemy positions more often than not. This in itself was a crucially important finding. It was soon joined, however, by two rather less comfortable elements in the debate. Firstly there was the matter of racial characteristics and secondly there was what came to be called the 'flight to the front'.

In the eighteenth century there had been a feeling among tactical theorists that the further north one went in Europe, the less active and excitable would the local inhabitants appear. Their soldiers would be less well fitted for rapid manoeuvres and bayonet charges, but much better at the careful and steady business of firing musket volleys. Germans were thought to fight best by fire, while Frenchmen were better at charging and individual skirmishing.

These ideas had been present in the tactical debate for over a century, although admittedly the firepower of the Germans had not seemed particularly impressive in the Napoleonic wars. What gave the theory a new lease of life around 1859 seems to have been the increased general interest in 'heredity'. Horse breeding was starting to be more scientifically based at this time, and it is perhaps no coincidence that Darwin's *Origin of Species* first appeared in 1859. Throughout the second half of the nineteenth century a variety of pseudo-scientific racial theories gained wide circulation in Europe, and they inevitably left their imprint upon the discussion of tactics:

The French, the Italians, and Latin armies in general have . . . a more ardent temperament than the troops of Germanic stock. As a result firing, which especially calls for *sang-froid* to be effective, is less suitable for them and pleases them less than the bayonet attack or individual action, in which their fiery natures can be given free rein.[28]

Having thus 'scientifically' identified the tactics which were natural for the French, theorists could believe that great advantages might be gained by taking them to extremes.

Attempts by officers to restrain the racially 'natural' ardour of their troops, or to stick to some rigid drill-book, could be represented as disadvantageous. It was therefore with some glee that the analysts of 1859 reported that the officers had sometimes been left behind in the charge:

The bayonet was in fashion in the whole of the allied army. It was a veritable frenzy, and very often the officers, far from commanding their men, were forced to obey those of them who threw themselves forward with cries of 'à la baionette!', and follow them whether they liked it or not. Before Melegnano, at the cemetery of Solferino and beneath San Martino, these courageous impulsions admittedly cost us dear; but the final result crowned them with success.[29]

This 'flight to the front' was also recognised as being partly the result of the improved firearms themselves. With bullets falling around a unit at long range, the men would be goaded into advancing all the faster. Paradoxically it was greater defensive firepower which actually speeded up the attack and gave it more enthusiasm. Losses to fire might be heavy; but they could be minimised if the soldiers did not stand still as passive targets.

Some commentators, however, saw that because it was fear which drove men forward, their attack would not be solidly based. It would lack the deliberation and cohesion which Bugeaud had recommended, and would be fragile if it encountered difficulties such as an enemy counter-charge. According to the members of this school the best tactics would be those which could somehow reduce the disorder and dispersion which the modern battlefield seemed to impose. Colonel Ardant du Picq in particular made a remarkable collection of eyewitness accounts of battle, which convinced him that the history of modern warfare was in reality the history of increasing fear and confusion. As firepower improved, so the ability of soldiers to retain their calmness decreased. It was thus but a short logical step to the view that measures should be taken to increase calmness and discipline, in order to reduce the effects of fear:

Combat requires today, in order to give the best results, a moral cohesion, a unity more binding than at any other time. It is as true as it is clear, that, if one does not wish bonds to break, one must make them elastic in order to strengthen them.[30]

Ardant du Picq, himself a former Chasseur, was advocating a looser but stronger form of discipline which would allow troops to fight more independently but more effectively in the dispersed and somewhat disorganised formations of the modern battlefield. Unlike Bugeaud or Napoleon III he did not feel that close order was necessarily the best answer to the problem, but he certainly did recognise the need for what he called 'order'. 'The victor will be he who secures most order and determined dash.'[31]

Good order alone impresses the enemy in an attack, for it indicates real determination. That is why it is necessary to secure good order and retain it until the very last. It is unwise to take the running step prematurely, because you become a flock of sheep and leave so many men behind that you will not reach your objective.[32]

The new discipline which was required would be more individual and internal than before. It would be what the Germans were later to call 'Innerführung' or 'inner leadership'. This would replace the more physical supervision of the past.

Today the soldier is often unknown to his comrades. He is lost in the smoke, the dispersion, the confusion of battle. He seems to fight alone. Unity is no longer insured by mutual surveillance. A man falls, and disappears. Who knows whether it was a bullet or the fear of advancing that struck him![33]

In order to retain his value as a fighting man, each soldier would henceforth require much greater personal commitment and indoctrination.

During the late nineteenth century it at least came to be accepted that infantry would no longer fight in closed formations, but in more or less open skirmish lines. Skirmishing would take longer and have greater importance for the final result. Supporting columns would become much lighter than they had been in the past, and their function would be to feed the skirmish line rather than to make the final attack directly. A final attack of some sort would nevertheless remain essential, and it would be delivered with the bayonet. The 'battle of killing' was not to be allowed to dominate the scene, since a decisive charge would always remain the ultimate arbiter. The important thing was to keep the men in a sufficient state of enthusiasm and freshness to make such a charge.

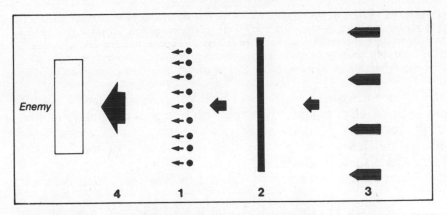

Late nineteenth-century infantry attack – building up the skirmish line on the 'empty battlefield'.

1 *Enemy engaged by skirmishers.*
2 *Line of supports to feed the firing line.*
3 *Small columns to feed in reserves gradually.*
4 *Attack with bayonet once the line is strong enough to win fire superiority.*

In essence the above was the tactical doctrine adopted by every country during the fifty years before the First World War. The idea was to build up a heavy skirmish line which could wear down the enemy and finally make a bayonet charge. The line would be held together by the training and individual determination of each soldier, who would have been indoctrinated in the need to keep moving forward. Everything possible would have been done to increase his 'morale' and to harness his supposed national characteristics.

At different times and in different nations there were minor variations in the details of these tactics, but it was rare for their basic outline to be changed. Around 1867–87 in France and around 1902–9 in Britain there did seem for a time to be a willingness to fight entirely by fire.[34] In these two cases the final bayonet charge dropped temporarily out of official doctrine, to be replaced by a conception of a genuinely 'empty' battlefield. In neither case, however, did such a phase last very long. It was soon overtaken by a return to the policy of the bayonet charge and high offensive morale.

Time after time it was found that in practice the bayonet charge could actually work. In the Austro-Prussian war of 1866, for example, the Prussians adopted a type of attack by self-sufficient fractions which must have been very close to what Ardant du Picq recommended:

67

Like wasps, when their nest is disturbed by some blunderer, they swarmed all round their dazed enemy, and put him to flight by the countless attacks of small groups converging at the right time and place. They received no orders to do so any more than the Austrians, but, from Subaltern to General a thorough military education had developed in them the reflexes necessary for the proper exercise of command in war.[35]

In the Franco-Prussian War it was certainly found that infantry assaults could be prohibitively expensive, even with troops of exemplary morale. The Prussian Guard attack at St. Privat, in particular, cost 8,000 casualties and left a deep scar on the consciousness of the European military establishment. On the other hand it would be wrong to suggest that infantry attacks were always doomed to fail, even when delivered frontally.

Prince Kraft zu Hohenlohe Ingelfingen has described a number of successful attacks which took place in the Franco-Prussian War. At Le Bourget on 30th October 1870, for example, the Prussians stormed two villages over perfectly open ground with trifling losses. In one case it was actually done without any support from the artillery:

At this point there were two battalions of the 'Franz' regiment who had to attack over 2,000 paces of open ground. The officer commanding this regiment had already practised it in the attack. In accordance with his practice he sent forward the whole of the leading line, which consisted of two companies, in thick swarms of skirmishers, and made them advance over the open ground in two parts (by wings) which alternately ran in 300 paces. After each rush the whole of the wing which made it, threw itself down, and found some cover among the high potatoes; there they recovered their breath while the other wing rushed in. As soon as they arrived within range of the needle-gun, the wing which was lying down opened a fire of skirmishers on that edge of the village which they were attacking. I can still remember, as I write, the delight which we felt as from our position we watched this attack which had been so carefully thought out, and was so well carried through. The best of the thing was that, as the commander of the regiment assured me, these troops suffered no loss up to the time when they reached the edge of the village.[36]

With examples like the above before them, infantry commanders after 1870 can have seen no reason to suppose that modern firepower had really made the attack obsolete. In both France and Prussia, indeed, there was actually a tendency to place increased stress upon the offensive in the training manuals. Especially in France it was felt that excessive caution had been displayed during the war, sometimes with dire results. The very defeat of the Prussian Guard at St. Privat, for example, was a case in point. On that occasion the French had failed to exploit their success with a counter-attack, thus 'snatching defeat from the jaws of victory'. French writers after 1870 were determined that such lack of resolution should not reappear in any future war.

In 1877 the Russian army attacked the Turks, who had just been equipped with an excellent new breech-loading rifle. Although the Turkish forces were largely composed of raw conscripts, they exhibited an admirable tenacity in holding defensive positions and blazing away at any Russian who came in range. The Russian General Todleben later wrote that 'Such a shower of lead as that with which the Turks hail our troops has never before been employed as a mode of warfare by any European army.'[37] Western observers also felt that this was something new, and once again the cry went up that firepower had finally made the bayonet attack impossible.

The war correspondent of the *Daily News* said that in his opinion:

. . . the whole system of attack upon even the simplest trenches will have to be completely changed in the future. Assaults, properly speaking, will have to be abandoned.[38]

He felt that only two kinds of infantry action would henceforth be possible – wide envelopments which avoided frontal attacks, or steady sapping forward towards the enemy with the spade.

Events at the siege of Plevna, however, did not finally justify such complete pessimism. Those who wished to preserve the infantry attack were able to find a number of replies to their critics. As they had in the American Civil War, Western European commentators could show that the two armies at Plevna had been less than excellent in quality. The Russians seemed to have learned little since the Crimean War: their tactics were still ponderous, inflexible, and made inadequate provision for skirmishing. If their heavy columns were shot to pieces, it was scarcely surprising.

It was also true that in many cases at Plevna the bayonet attack had achieved a measure of success. Time and again it succeeded in capturing the front line of enemy trenches, thus allowing the attackers to install themselves at close range for a subsequent assault. The true failure was seen as a failure to maintain the momentum which this initial attack had created. It was not so much 'the bayonet attack' which was difficult, but rather 'the second bayonet attack' with blown troops. If methods could be developed to deliver a quick exploitation of the first success, then infantry offensives would still be quite possible.

In practice one Russian General, Skobelev, did succeed in devising a method for ensuring a rapid renewal of impetus at the moment when it was needed most. This consisted of committing a new echelon of attacking infantry at the moment when the first wave started to falter. A great deal depended upon good judgment of the psychological moment, and it required the overall commander to be posted well forward where he could see what was going on. In practice, however, Skobelev made the system work in his attack on the Kazanlik redoubts. This was a very costly attack, it is true; but it did reach its objective. Many commentators felt that it had important lessons for the future of infantry.

There was unquestionably a great deal of genuine bayonet fighting at

Plevna, 'a pure chaos of stabbing, clubbing, hacking, clutching, shouting, screaming men', as a participant described one such combat.[39] There were also many occasions when a defender ran away in front of a threatening charge, before it made contact. We must further remember that this battle was generally considered to be exceptional by observers, insofar as it took on the characteristics of a siege. The general feeling was that the Russians would have done a great deal better if they had stuck to open, mobile warfare. On other fronts of the same war they had indeed done so with a much better outcome.

It is only too easy for modern writers, with full benefit of hindsight, to assume that the development of firepower in the American Civil War, in the Franco-Prussian War and at Plevna ought to have made frontal attacks unthinkable. The fact of the matter is that contemporaries did ponder the problem at length, but concluded on the basis of the evidence that such attacks could still be made. All they required was a little extra care, and an acceptance of increased dispersion in the skirmish line.

The most triumphant vindication of this view seemed to be the Russo-Japanese War of 1904–5. Coming as it did immediately after the Boer War, observers had expected to find an even more empty battlefield, with more killing and less decisive results. Instead, they were astonished to find a series of actions similar to the following:

It was about 3.45 of the clock when the brave Japanese broke cover in one long line and headed due north. The men were almost shoulder to shoulder in single rank. The supports followed at about 200 yards, also in single rank, and behind them came the reserve in double rank. There was no firing. The rank and file marched with sloped arms and fixed bayonets and swung along steadily, almost solemnly, forward . . . Okasaki's infantry . . . reached Kokashi without firing a shot themselves, and without, so far as I could see suffering any loss at all from the bullets which had been raising little puffs and spurts of dust about them as they advanced. As for the Russian guns, they had either failed to detect the commencement of the attack, or else they had been successfully distracted from their legitimate target by the Japanese bombardment.

It was soon after four o'clock that the Brigade Okasaki disappeared into Kokashi and into a village half a mile to the east of it called West Sankashi. Then there arose a continuous tearing crepitating sound, not very loud, and yet sufficient in intensity and volume to cause us all to shiver with excitement. . . . It startled me like the sudden snarl of a wild beast. For I knew that thousands of rifles had opened magazine fire and were struggling at from 500 to 600 yards distance for the fire mastery, that fire mastery which, established by the one side would render the assault possible; established by the other must doom it to disastrous failure. Such sounds as these, wafted upon the evening breeze, bore messages of life and death; more – of victory or defeat to all who could grasp their significance.

For a long, long time the anguish of anticipation was spun out to the uttermost. A quarter of an hour passed, then another quarter of an hour; the General Staff could hardly endure it any longer, but Kuroki (the Commander-in-Chief) remained confident and calm. Then another ten minutes. The tension became unendurable. . . . 'Ah,' said Kuroki, 'He cannot get on. Today we are stuck fast all along the line.' In his voice was no tone of regret, no shade of mortification; at the most it could only be said that the actual words betokened some touch of despondency.

Hardly had he spoken when a sharp exclamation from an Adjutant made me turn my glasses once more upon the deserted plain, and to my amazement I saw it, deserted no longer, but covered by a vast, straggling, scattered crowd of individuals, each racing towards the Russians at his topmost speed. The Okasaki Brigade was crossing the open to try and storm Terayama by one supreme effort; and the only English expression which will convey an idea of their haste is that of the hunting-field, 'Hell for leather'. Bullets fell thick among those who ran for life or death across the plain, and the yellow dust of their impact on the plough rose in a cloud almost up to the men's knees. By what magic these bullets almost always struck in the vacant spaces and very rarely on the bodies of the men, I cannot explain, beyond saying that it was ever thus with the bullets of a bad shooting corps. At the first glance it seemed as if there was no order or arrangement in this charge of a brigade over 500 or 600 yards of open plough. But suddenly I realised that it was not chance but skill which had distributed the pawns so evenly over the chess board. The crowd, apparently so irregular and so loosely knit together, consisted of great numbers of sections and half-sections and groups working independently, but holding well together, each in one little line under its own officer or non-commissioned officer. There was no regular interval. . . .

In certain respects the startling, sudden onslaught of Okasaki's Brigade resembled a Dervish rush, but with one marked difference, inasmuch as the formation was not solid but exceedingly flexible and loose, offering no very valuable target even to a machine gun. The speed was marvellous, and the men got across the plain more like charging cavalry than ordinary infantry. Some say that the leading sections paused once to fire. I did not see this happen. To the best of my observation the assaulting infantry ran 600 yards without the semblance of a halt, as their leading files reached the sunken road they dashed unhesitatingly into it, right onto the top of the crouching Russian infantry! Next second the Russians and their assailants were rushing up Terayama slopes in one confused mob, the whole mass convulsively working bayonet and bullet and clubbed rifle as they ran. The hill was carried. Bravo! Bravo!! Bravo!!![40]

Even making allowances for Sir Ian Hamilton's obvious emotional involvement, it is clear that Okasaki's charge constituted a serious reproach for those who believed such things were impossible in the age of the magazine rifle. It is true that Hamilton elsewhere lists a number of factors which greatly helped the Japanese, for example the bad Russian marksmanship and their poor artillery co-ordination. Nevertheless the very fact that such attacks succeeded at all was more than sufficient to keep the idea of the bayonet alive.

There were really two wars in Manchuria in 1904–5. The first was the mobile war which Sir Ian Hamilton reported so vividly. The second was the positional war of Port Arthur and Mukden. It is the latter which in retrospect seems to have foreshadowed the Western Front so closely; but it was the former which attracted much of the attention at the time. It was easy for analysts to dismiss the sustained trench-fighting as 'sieges', in the same way as they had done for Plevna, and indeed for Petersburg in the American Civil War or Sebastopol in the Crimea. Military operations had always been divisible into the two general categories of 'sieges' and 'open warfare'; so it surprised nobody when the same pattern was repeated after the introduction of magazine rifles. 'Sieges are horrible things,' said Hamilton. 'A good fight in the open – that is another matter.'[41]

It is also very easy for us to forget that until the start of 1915 it looked very much as though the First World War had itself taken on a highly mobile character. In the early months of war there were some astonishing demonstrations of the power of the offensive. At Tannenburg the Germans manoeuvred between two superior Russian forces and destroyed them both. At Lemburg the Russians in turn effected a breakthrough of the Austrian positions. In the West the Germans had quickly defeated the Belgians in 'siege' warfare at Liège, and had then advanced almost a hundred miles into France, pushing all before them. At the Battle of the Marne it was the turn of the counter-offensive to achieve some startling results. To contemporaries all this must have seemed a veritable firework display of offensive action. If not even the Belgian fortresses could hold out for many days, what hope was there for the defensive in general?

The establishment of a line of trenches from Switzerland to the Channel admittedly represented a pause in the offensive impetus; but even the most doctrinaire bayonet-fanciers had always admitted that some sort of let-up would sooner or later be inevitable. There seemed to be no reason to doubt that in the spring of 1915 the deadlock would be broken and the offensive would again be set in motion. If there was now a momentary 'siege' phase in operations, it was presumably the result of a temporary exhaustion of both logistic and moral assets. As soon as these had been made good the armies would surely be free to return to 'a good fight in the open'.

Field Marshal Rommel has left us some startling testimony of just how free from casualties the offensive could be in 1914, when he was still only a Platoon Commander. Advancing on a French village in morning fog, he reports that:

Suddenly a volley was fired at us from close range. We hit the dirt and lay concealed among the potato vines. Later volleys passed high over our heads. I searched the terrain with my glasses but found no enemy. Since he obviously could not be far away, I rushed towards him with the platoon. But the French got away before we had a chance to see him . . . several additional volleys were fired at the platoon from out of the fog; but each time we charged the enemy withdrew hastily. We then proceeded about a half mile without further trouble . . . . So far the platoon had suffered no casualties.[42]

Again, in another attack on the same day he reports that:

We rushed forward by groups, each being mutually supported by the others, a manoeuvre we had practised frequently during peacetime. We crossed a depression which was defiladed from the enemy's fire. Soon I had nearly the whole platoon together in the dead angle on the opposite slope. Thanks to poor enemy marksmanship, we had suffered no casualties up to this time. With fixed bayonets, we worked our way up the

72

rise to within storming distance of the hostile position. During this movement the enemy's fire did not trouble us, for it passed high over us toward those portions of the platoon that were still a considerable distance behind us. Suddenly, the enemy's fire ceased entirely. Wondering if he was preparing to rush us, we assaulted his position but, except for a few dead, found it deserted.[43]

Finally, Rommel makes an interesting observation about the power of modern artillery:

. . . The whole clearing was subjected to intense shrapnel fire. The shells rained down like a sudden thunderstorm. We tried to find shelter behind trees and used our packs to form improvised breastworks. The intensity of the bombardment made it impossible to move in any direction. Although the bombardment lasted several minutes, there were no casualties. Our packs intercepted a few of the missiles, and the bayonet tassel of one of the men was torn in shreds.[44]

The above passages reveal precisely how innocuous the much-vaunted 'modern weapons' of 1914 could actually be. There did not seem, by and large, to be any reason for thinking that a great revolution had suddenly taken place. There seemed to be plenty of justification for the continuity in tactical affairs which had generally been assumed before the war. The British commentator Spenser Wilkinson had summed up this feeling in 1891, when he observed that:

It is true that within certain narrow limits, which can be precisely specified, the defender is strengthened by modern improvements in firearms. But it is not true that this results in a great or sudden change in the relations of attack and defence, either in regard to battle as a whole, or in regard to the general course of a campaign. There has been no revolution in tactics or in strategy, but certain modifications long since realised have become more pronounced. The balance of advantage remains where it was.[45]

Spencer Wilkinson seems to have grasped a vital truth about this tactical debate which many modern commentators have totally missed. At no point during the nineteenth century *did* anything very much seem to change. It was always accepted that steady improvements in firepower would be going on all the time, but that these would be balanced by a gradual lightening of the fighting line and a gradual improvement in the sciences of leadership and morale building. From the very start of the century to the very end, some prophets of doom could always be found who believed that the bayonet charge was finished. The rifle enthusiasts of the Napoleonic wars had seemed to suggest as much, as had the Chasseurs of the 1830s. Others had said the same after the wars of the 1860s and 1870. Plevna, the Boer War and Mukden had all seemed to prove the point yet again. But for all that, the bayonet charge did nevertheless continue to work in practice.

An important feature of the bayonet charge was that there was very little real alternative. If one were to follow the advice of the *Daily News* correspondent at Plevna, for example, one would have used either a wide outflanking movement or a prolonged approach by trench warfare. Neither option was particularly inviting, even though the idea of encirclement did actually gain a great deal of support from the time of the American Civil War onwards. The Prussians were especially fond of this tactic during the Franco-Prussian war, and it later lay behind their great Schlieffen Plan itself. The British also found by their experience in South Africa that some form of encirclement could be very helpful indeed, and the Japanese used it frequently in Manchuria.

Encirclement of the enemy did not take as long as sapping forward; but it still took quite long enough, and exposed the encircling forces to a loss of contact with their main body. It also traditionally required a great superiority of numbers as well as an enemy who would obligingly stand and watch while you rolled him into your trap.

All of these considerations meant that encirclement was frequently either impractical or very risky. For a much quicker and apparently more certain solution the odds often seemed to favour a frontal attack.

Once a frontal attack had started, the choice was between a 'battle of killing' and a genuine assault to close quarters. Here again the choice would be highly unattractive, although commanders could not help but feel that a 'battle of killing' would mean precisely what it said. For example Hamilton reports one battle in which the Japanese were faced with just this problem:

> . . . it was quickly realised by regimental officers and men that the fire was too hot to admit of a prolonged duel between troops in the open and troops under cover, and that the only alternative to going back was to go forward. Instinctively the whole line endeavoured to press on.[46]

Here we again find the logic of the 'flight to the front' as described by Ardant du Picq and his school. It did not seem to be possible to stand still under fire, and it would certainly not have been decisive. If men were going to be hit anyway, then they might just as well have achieved something positive to compensate for their sacrifice.

Behind the tactical assumptions carried into the First World War there lay a great deal of sound reasoning, based upon both concrete examples and abstract principles. Repeated attempts had been made to challenge this reasoning, but the requisite body of proof had always been found wanting. Every time firepower had seemed to make some great stride forward, the power of the bayonet had somehow managed to keep pace. It was only impractical visionaries who would try to deny this and suggest that improved firepower had at last made no-man's-land genuinely impassable. There really did not seem any good reason to listen to men like Morand with his steam machine-guns of 1829, or to the Polish financier I. S. Bloch with his highly abstract arguments of 1897.[47]

The case for the other side was very strong indeed, and at the time it seemed to be even stronger.

Today we like to think of late nineteenth-century bayonet enthusiasts, such as de Grandmaison or Dragomirov, as being little more than dangerous cranks. By stressing one particular element in war they certainly sacrificed the overall balance of their views; but in the perspective of their own day they were quite right to choose the offensive as the element which needed to be stressed. This was the part of war which had been made more difficult by improving technology, and yet it was also an unavoidable necessity imposed on any Commander. In these circumstances it was logical to find a counter-balance to firepower in the shape of morale and the spirit of aggression. These things were apparently solidly based in the psychological and racial sciences of the day, and had been analysed in a century of careful debate. They were known and trustworthy instruments which seemed to be increasing in power as much as the new weapons themselves.

In the First World War it was not only cannons and machine-guns which finally did prevent movement across no-man's-land; it was also the extraordinarily strong morale which had been built up in every army. It took three years for the armies to crack, and even then some of them never did. To this extent, at least, the pre-war interest in morale had not been in vain.

The Alleged Rifle Revolution in the American Civil War

Within their own terms of reference the European armies had prepared themselves well for the First World War. They had examined new technologies such as the magazine rifle or the 'Aerial Torpedo' (the HE shell), but they did not believe these would radically change the shape of the battlefield. Believers in firepower perceived a continuum of infantry tactics from Frederick's drilled 'automata', through Wellington's allegedly steadily-volleying regiments, up to the present day. Since infantry had always fought by massed fire, the Maxim gun was merely doing mechanically what a platoon of Potsdam grenadiers would previously have done manually. Conversely the believers in morale and the offensive could trace a no less respectable continuum from Napoleon's 1805 column attacks covered by skirmishers, to the 1905 Japanese bayonet dashes prepared by musketry. If technology had made preparatory fire a little more important to an assailant than it had been in the past, that did not seem to make a very decisive difference to the time-honoured relationship between attack and defence.

Perhaps one important reason for this complacency was the need for

75

an inter-arm balance in the debate. If the cavalry could still see a chance to fire on foot and then charge mounted, the artillery could just as easily imagine a battle in which their shells destroyed literally everything. If the infantry believed they could fight their own way forward by fire and movement, the engineers were thinking more about creating impenetrable defences with mines, electrified barbed wire and deep, concrete-lined bunkers. Imagination was free during the forty years since the last major European war, and because there was little central co-ordination to the debate, its constituent parts tended to cancel each other out. No one could tell just where the actual balance should be struck, so they carried on behaving normally.

The shock by the start of 1915, therefore, was not so much a realisation that new elements had entered the nature of combat, but that one side of the pre-war debate had apparently been proved right and the other wrong. The debate reached a clear turning point, since for the first time in a century there seemed to be a new style of warfare that almost everyone could agree was radically different from that of the past. Maybe this was really true: it makes an interesting question to which we will return in the next chapter. Suffice it to say here, however, that a decade or so after the Treaty of Versailles, most armies were looking forward to fighting their next war in a novel manner compared to what had gone before.

We have already seen how a succession of prophets had prematurely announced this revolution at several different moments during the nineteenth century. 'The new warfare' had been heralded, variously in the Crimea, in 1870, and at Plevna. Manchuria had impressed more commentators than usual, and, in Britain at least, the Boer War had made a profound impression.[48] None of these claimed revolutions had really stuck in the public consciousness, however, and traditionalists were reassured to think that the battle of Omdurman, as late as 1898, could still be won by cavalry charges and infantry squares.[49]

It was only in one part of the world that a belief in the arrival of 'the new warfare' had become deeply rooted before 1915, and survives to the present day. This was in America, where the novelty of the Civil War of 1861–5 impressed the population even more deeply than the Franco-Prussian War had impressed the French. In France the 1870 war was settled by a couple of 'Napoleonic-style' lightning campaigns, whereas in America the armies seemed to lack the tactical ability to reach clear-cut decisions on the battlefield, and fighting dragged on for four years. Worse, the indecisiveness of the battles meant that the North's victory could be won only by a systematic attack on the Confederate economy. By naval blockade, mounted *chevauchée* and scorched-earth marauding, the Southern civilians were made to pay for their rebellion. Nothing quite comparable was seen in the European wars of the Victorian era, although more bitter scenes had certainly featured in the campaigns of both Wellington and Napoleon.

The Civil War could obviously be seen as a break with the traditions of the past. Before First Manassas in 1861 the USA had never put more than 20,000 men into a battle, but by the end of the Civil War over three million Americans had passed through the armies, and many battles had been fought with over 30,000 men on each side.[50] Before the 1860s the Americans may have read much about Napoleon's exploits, but they had perhaps paid more attention to his glitter and glory than to the stern demands he had made upon his subjects. It must have come as a major trauma to find that the Civil War cost a third of a million Union and Confederate dead (to all causes, including disease), and maybe half a million wounded.

If these novelties had been the only ones to strike the American public during the Civil War, they would probably still have been acclaimed as 'the birth of a new warfare', even though they contained nothing particularly new to Europeans. However, there were several other technical factors at work as well, which went at least some way towards persuading even the Europeans that something new might really be afoot. There was the first truly widespread and sustained use of steam power in war – a massive expansion on Jackson's early use of a steam launch at New Orleans in 1815. There were iron-clad warships, observation balloons, and 'torpedoes' (hidden, percussion-detonated mines for both land and sea); and a number of prototype terror weapons ranging from machine guns to exploding bullets, and from railway mortars to poison gas. Making a particularly strong impact upon the general public, there was a recognisably modern newspaper coverage of the campaigns, including biting satire against many of the generals and a lovingly gruesome emphasis on the dead, the dying and the bereaved. All this meant that the war not only felt new by reason of its duration and scale of privations, but it also looked new in its machinery and presentation.

The cornerstone of the Civil War's supposed modernity, however, consisted of the new rifles and cannons deployed in battle, which were widely assumed to have changed the whole nature of tactics. By 1865 every infantryman carried a rifle sighted to around 1,000 yards; most of the cavalry had repeating pistols or carbines, and a considerable proportion of the artillery was also rifled. In theory this made a huge increase in firepower, and many are the tactical analysts, both past and present, who have produced their own version of the following sentiment:

The introduction of the rifle musket and its conoidal bullet in the decade between 1850 and 1860 was to have the greatest immediate and measurable revolutionary impact on war of any new weapon or technological development of war before or since. When and if tactical nuclear weapons appear on the battlefield, presumably they will have an even greater effect. But certainly not even the high-explosive shells, airplanes or tanks of the twentieth century were to have effects of contemporary scale and significance comparable to the rifled musket in its early days.[51]

It is certainly hard to see that the British use of the Minié rifle at Sebastopol had an effect on the musket-armed Russians comparable to an attack by tanks and planes, or that the Austrian Lorenz rifles at Solferino could put down a wall of fire equivalent to the HE bombardments of the Somme or Passchendaele. Nevertheless, the impression remains strong in many people's minds that the Springfield and Enfield rifle muskets carried by Civil War soldiers made for devastatingly more lethal scales of firepower than anything seen before.

It is not difficult to show that this assumption is totally misleading,[52] since many of the Civil War battles were fought in close country where long fields of fire could not often be found. The great range of the new rifled weapons could not usually be exploited even if the troops had been suitably trained, which they generally were not. Whereas in the Zulu War of 1879 we hear of range markers being laid out to 700 yards from a infantry position,[53] there seems to have been little comparable to this in the Civil War, where the normal exhortation, as in the past, was simply to wait until close range and then aim low. When fields of fire were specially cleared, they usually extended no more than 50–100 yards. Eyewitness accounts seem to suggest that in combat the average range for musketry in the war as a whole was 127 yards, and 141 yards for 1864–5. Out of a sample of 113 references to range, the author found 70 (or 62%) showing fire at 100 yards or less, 96 (or 85%) at 250 yards or less, and 100% at 500 yards or less.[54]

Although these average ranges are considerably longer than those in Wellington's infantry fights, they are probably not so very much greater than those familiar to continental Napoleonic armies, and generally not vastly incompatible with Napoleonic expectations overall. There had been no marked increase in the rate of fire with rifle muskets over that with smoothbores, so the total number of shots that could be fired at an attacker before he could reach a defender was surely not much greater. There is also a strong impression, from many accounts of Civil War combat, that after only a few opening shots at these ranges the two sides would approach closer to each other for a static exchange of fire. This might last a long period of time – possibly until the full load of forty or more rounds had been expended by each man. Hence only a small proportion of the total bullets fired (and therefore casualties inflicted) would be at the longer ranges, even though those ranges themselves might exceed Napoleonic norms. It is probably worth adding that due to the smallness of the aiming mark he can see, regardless of the qualities of his weapon, the average soldier under the stress of combat in any era is incapable of very much accuracy with a shoulder arm at ranges greater than the Napoleonic 'fifty or a hundred yards'. In most wars it will therefore be within that zone that the majority of the killing takes place, and it seems that the Civil War was no exception.

If the range and rate of fire of Civil War musketry showed little change from Napoleonic times, the same was also true of combat formations. The chief drill manual of the war was Hardee's translation of the French *chasseur*-inspired tactics of the 1850s, which advocated formed bodies two deep moving at double quick time, and also provided encouragement to the skirmisher.[55] These tactics could perhaps be claimed as an early response to the battlefield problem of heavy rifle fire; but it is clear that their full gymnastic rigours were properly applied in America no better than they were in Europe. On the contrary, American tacticians often seemed to revert to the slow and heavy columns, or succession of lines, that Jomini and Bugeaud had argued against, and which the *chasseurs* had been specifically designed to avoid. Time and again in the Civil War an enormous mass of troops was herded forward on a very narrow frontage, leading to notorious 'slaughter pens', such as Marye's Heights at Fredericksburg, Pickett's charge at Gettysburg, or the Union attacks at Kenesaw Mountain and Cold Harbor. These were the same tactics that had often been discredited against Wellington in the Peninsula, no less than against Andrew Jackson at New Orleans. It was pretty inevitable that they would meet the same fate in the 1860s as they had fifty years earlier, and we need not look for any new power of weaponry in order to understand the results.[56]

Civil War generals do not appear to have been very interested in minor tactics, and set up no authoritative body to analyse and develop doctrine, or to create élites that might specialise in assault techniques. This meant that battles were usually fought with somewhat rudimentary tactical concepts, leading to toe-to-toe firefights where both sides hoped to win by firepower, but where the defender was usually better prepared and held most of the trumps. Whether or not he used fieldworks, the general atmosphere of these encounters is well caught in the following passage:

> . . . it was a stand-up combat, dogged and unflinching, in a field almost bare. There were no wounds from spent balls; the confronting lines looked into each other's faces at deadly range, less than one hundred yards apart, and they stood as immovable as the painted heroes in a battle-piece . . . and although they could not advance, they would not retire. There was some discipline in this, but there was much more of true valor.
>
> 'In this fight there was no manoeuvering, and very little tactics – it was a question of endurance, and both endured.[57]

This in itself helped to make tactics less brilliant than they might otherwise have been, but the problem was compounded by failures in higher battle-handling. All too many Civil War commanders found they lacked the conviction, the staffwork, or the understanding necessary to convert a local victory into a break-through and pursuit at the grand tactical or 'operational' level. Despite the many difficulties involved in 'breaking into' an enemy position, it could sometimes be done; but there was a further crucial problem of generalship when it came to 'breaking out'. In the absence of a reliable method of achieving this, each local success seemed merely to increase the list of casualties without hastening the end of the war.[58]

By 1864 the Civil War soldiers had grown weary of throwing themselves into slaughter pens which created no armistice, so wherever possible they tried to apply more elusive tactics – elusive primarily from the enemy's fire, but also from their own officers. This was an era when the armies had advanced beyond the status of 'veteran' to that of 'old lag', wise in the ways of the system. It was an era of 'shadow boxing', when regiments ordered to make an attack might content themselves with advancing a few yards and firing for two or three hours, and whole armies might manoeuvre skilfully against each other, but decline close combat when they had the chance.[59] In the Eastern theatre the non-battle on the Mine Run stands as a classic example of this 'live and let live system',[60] while in the West the near-platonic minuets of Sherman and Johnston, between Chattanooga and Kenesaw, are even more famous. It took a very strong-willed commander, such as Hood at Atlanta or Grant in the Wilderness, to persuade the armies to return to their previous assault doctrines; but when they did so it was unpopular with the troops and led merely to increased human losses without compensating tactical gains.

Many commentators have seen the trend to 'live and let live' in 1864 as a move away from European formalism and towards 'Indian' or 'American' skirmish tactics. This would be convincing if the skirmish lines of the period were able to make co-ordinated assaults in the manner of the 1870 'Prussian Rush' demonstrated at Le Bourget in 1870, but there is little evidence that they were ever intended to do so. The few successful skirmish attacks that can be found are usually at night or in mist: two conditions that were generally neglected until at least the First World War. In the Civil War the skirmish line was regarded as an auxiliary or standoff means of fighting.[61] For serious assault work a heavier formation was preferred, right up to the end. General Upton at Spotsylvania refined the concept by training his men in follow-up action for each element in the assault, but even he was still using a reinforced brigade column to beat down the enemy's earthworks. In his drill manual written after the war, on the basis of the war's lessons,[62] he still expects troops to fight mainly in double ranks, reserving fire to 'deadly range' in a distinctly Napoleonic manner.

Apart from standoff skirmishing, the other element of the 'live and let live system' which the Americans embraced with enthusiasm was the use of fieldworks. Like Andrew Jackson at New Orleans, they believed their inexperienced armies were weak at manoeuvres in the open, and needed the additional stiffening of entrenchments. General McClellan had seen the siege of Sebastopol, and leapt to the conclusion that it showed the shape of wars to come. General Halleck had also argued for something similar in his influential pre-war handbook,[63] and his advice was applied from the very start of the war, even when inappropriate. This excessive respect for fortification was as much the result of theoretical book-learning by West Point graduates – who were all expert engineers – as it was a practical adaptation to the tactical system adopted by the armies by 1864. A vicious spiral set in, whereby the advanced tactics of mobility and counter-attack were progressively neglected, but the elementary tactics of static fire and mud-digging became universal. Many commentators have wrongly attributed the emphasis on fortification to the power of new weapons, whereas it probably owed more to a failure of creative tactical thinking.[64]

Civil War infantry fighting does not seem to have progressed far beyond the level reached at New Orleans. However, those who seek a 'revolution' in the 1860s claim that at least the infantry rifles could now outrange artillery. The whole relationship between infantry and artillery is deemed to have altered in favour of the former, and Union casualty statistics from the Wilderness battle are often quoted to show that cannon now caused only 9% or less of the casualties.[65] This, alas, is a classic case of drawing too wide a conclusion from too narrow a data base, since the Wilderness battle was recognised by contemporaries as the one least favourable of all to artillery. General Grant sent home over 100 of his guns from that battle, and found most of the rest of them were confined to roads in thick woodland.

We must remember that at Malvern Hill the Union artillery is credited with 50% of the Confederate casualties, and at Gettysburg it was the most effective instrument in the destruction of Pickett's charge. Nor is there any evidence that Civil War infantry had become more adept than Wellington's troops at chasing off enemy field guns. In both cases it seems that isolated batteries were vulnerable when they came too close, but massed artillery supported by infantry was safe enough. Where guns were lost to infantry or cavalry, the gunners were usually overrun with cold steel rather than shot down from long range.[66]

Mention of the cavalry brings us to the one area that really did show a revolution in Civil War tactics. At the start of the war the cavalry was few in number, not always well led, and rarely given a role in pitched battles. It fulfilled its destiny merely as outposts, scouts or couriers, until the Confederate J. E. B. Stuart showed how it could grab both booty and glory by raiding away from the main army. This had the unfortunate effect of removing it still further from the pitched battles, and it was only in 1864, after the Union cavalry had been reformed, that a new main battle role was found. Sheridan, however, then led a cavalry corps for the Army of the Potomac which managed to combine operational mobility with high tactical firepower and shock. His men were equipped with repeating Henry or Spencer carbines, and would engage the enemy in a dismounted firefight before charging home on horseback with sabre or revolver. In the Appomattox campaign this system successfully loosened up the sluggish progress of the infantry and restored fluidity to operations.[67]

The value of Sheridan's carbines was based more on a high rate of fire and volume of fire than on precision or long range. As such, they leapt ahead a generation to the design qualities of the SMG and automatic assault rifle, rather than to the accurate, long-range but single-shot rifles of late nineteenth-century theory. They were so far ahead of their time that their true significance was widely missed – making perhaps the only solid foundation to the claim that the Americans of the 1860s were radically more advanced than their blind and effete European counterparts.

The mainstream of European development stayed with the long-range, single-shot rifle. Early models had been present in large numbers in the Civil War, but, as we have seen, their range was rarely exploited. The truly revolutionary development in this direction was to come in the Franco-Prussian War, where the fields were often much more open and the volume of fire could be much heavier. Whereas the Civil War rifle musket had still been a fairly clumsy muzzle loader, the *chassepôt* and the needle gun were both breech loaders. In theory, at least, they could manage six or ten rounds per minute, in place of one or two from the rifle musket. Especially with the *chassepôt*, sighted to 1,200 metres, a hail of lead could be maintained at long range by the use of 'coffee mill' fire – a process by which the infantryman used his left hand to hold the rifle to his right hip at 45° to the horizontal, and his right hand to load, fire and work the bolt as fast as he could.[68] The result was that even though long-range fire could be properly aimed no better than in any other war, it could now be made effective and dangerous in a way that it had not been in 1861–5. When a battalion's 'coffee mill' rifle fire was combined with the indiscriminate long-range fire of a *mitrailleuse* battery, it could make a vast beaten zone that enemy infantry would hesitate to cross. In battle it was often found that two opposing infantry lines would settle down to their static phase of fire when they were some 400 or even 1,000 metres apart.

Modern commentators have often incorrectly derided the contribution made by the *mitrailleuse* to the tactics of 1870,[69] but still less widely recognised today is the contribution made by the cavalry. It is true enough that a number of charges failed against infantry, especially those made by the French. Margueritte's famous final effort at Sedan, for example, was launched in a mood of suicidal despair rather than realistic hope of tactical gain, although even then his command remained sufficiently intact to charge a second time, a few minutes later.[70] Nevertheless, the German cavalry did sometimes make effective attacks, of which the most celebrated, but misunderstood, is the so-called 'death ride' of von Bredow's brigade at Rezonville. This has often been dismissed as another futile Balaklava – and the 'death ride' tag is itself more than expressive. Yet the truth is that two depleted cavalry regiments, totalling some 804 troopers, took a whole French army corps by surprise, overran its guns, paralysed its command functions for three hours, and prevented it from making an attack that could have been strategically decisive. They were finally beaten off only by French cavalry, not by the infantry or artillery. Admittedly the Germans suffered around 60% losses, but, as a cost-effective expenditure of casualties in order to win battles, their sacrifice can scarcely be faulted.[71]

Returning to the infantry, most armies were at this time equipping their troops with breech loading rifles. They believed that the newly increased volume of fire made genuinely long-range shooting worthwhile – provided that sufficient ammunition was available. The British, in the Zulu War of 1879, hoped to open fire at 1,000 yards or more with their Martini-Henry falling block rifles, although in practice it was often much less. However, the sheer volume of lead could usually hold the enemy at 150–300 yards' distance, and even if he successfully came closer it would normally only be in manageable numbers. At Isandhlwana, however, the British line was overwhelmed when the immediately available ammunition was exhausted, and the Zulus sensed that they had 'won the firefight'.[72]

By 1895 the British had digested these lessons, and with the Lee Metford magazine rifle – but also still with the Martini-Henry itself – they were advising against aimed fire at more than the 'decisive range' of 400–500 yards. Unaimed fire was acceptable at ranges up to nearly a mile, since a 'doctrine of chance' would ensure that at least a few of the shots hit their mark. However,

. . . it is a waste of ammunition, even if the range be known, for bad shots to fire *individually* at ranges beyond those for which the fixed sight can be used, and for good shots beyond 800 yards.[73]

In a French formulation of a somewhat similar point, de Grandmaison claimed that in Napoleonic times the crucial range had been 100 metres; in 1870 it had been 4–500 metres; with the M1874 Gras rifle it had been 5–600, and with the M1886 Lebel magazine rifle it was 800–1,000. De Maud'huy said that in 1800 infantry would break at 20 metres, but in

1910 at 200 metres.[74] A 1908 German view of defensive tactics advocated only sparing use of massed fire beyond 1,200 metres, but annihilating fire at medium range (800–1,200m) and short range (below 800m). The longer the range, the slower the rate should be; and skirmish fire was to be reserved for shorter ranges than massed fire. It was also claimed that every country wished its troops to hold their fire in the attack for as long as reasonably possible.[75]

All this is as remote from Wellington's aggressively mobile counter-attacks at 30 yards as it is from the largely unaimed Civil War volleys at 100 yards or less. In 1870 we had entered a new world of long-range fire, in which attacking infantry had to negotiate a much wider beaten zone than previously, before coming to the decisive final crossing of no man's land. The Americans' claim that they had seen the shape of modern warfare already in the 1860s, and had fully mastered it then, is surely wide of the mark. Nor was it to be confirmed by their subsequent tactical showing when they first came into action at San Juan hill in 1898, or Belleau Wood in 1918.

Pleasure Train to Berlin[76] – The Alleged Influence of Louis de Grandmaison on the French Army in August 1914

In their Civil War the Americans suspected they were entering a new age in warfare, not solely because 98% of their armies were new to the military life, but because futurist military analysts had prematurely announced a new age of rifles, trenches and infernal devices of every kind. In 1914, however, the French army set out to war expecting relatively little to have changed since 1870. The troops wore the same uniforms and were due to fight on very much the same battlefields, supported by a network of permanent fortifications which included some of Vauban's original works. Field artillery and rifles had not dramatically increased in effective battle range since 1870, although the incomparable rapid-firing *soixante-quinze* had replaced Napoleon III's cranky old muzzle loading cannon, and the *chassepôt* had found an eminently worthy successor in the Lebel magazine rifle.[77] Therefore, there was a strong feeling that the new battle would in fact be the second round in a continuing struggle, rather than a break with the warfare of the past.

Many lessons had been learned from 1870, of course. This time the new war was not entered with 'lightness of heart', but with a carefully studied 'reluctance' in face of overwhelming aggression. This posture secured powerful allies in place of diplomatic isolation, and the support of a mass citizen army rather than merely a small band of Bonapartist mercenaries. The rail-born mobilisation had also been carefully studied and ran like clockwork, bringing five French armies to their correct positions and on time – *quelle différence avec 70!*[78] Only the British arrived late and with fewer troops than they had promised: but what else could one expect from this perfidious neighbour? The BEF's mobilisation may have run smoothly according to its own lights, but it exasperated the French and led to a long trail of misunderstandings. Once fighting had started, the Mons position was held for only a few hours; only half the available troops fought at Le Cateau, and then Sir John French panicked and tried to make off to the ferry boats at St Nazaire![79]

Some of the lessons learned from 1870, however, were less than constructive – notably the new perception that all French territory was henceforth sacred. Bismarck's cynical attempt to allow France an African empire as compensation for the loss of Alsace-Lorraine was deeply unacceptable, since provinces could no longer be bartered away as they had been under the dynastic statecraft of the eighteenth and early nineteenth centuries. In the Third Republic the whole of France belonged to all its people; it was no longer the personal estate of some king or emperor. This was not merely a political imperative for a government which relied on the support of a nation in arms, but was also a matter of deep religious and racial concern.

The popular mind at this time was filled by the rantings of spurious experts who claimed pseudo-scientific support for the most garish forms of chauvinism, bigotry and prejudice. There was a hothouse atmosphere which produced both the expiatory Catholic pilgrimage movement of the 1870s, and the proto-fascist, anti-Dreyfusard Boulangism of the 1880s and 90s.[80] The cult of Joan of Arc was harnessed to the military crusade for *La Revanche* when it was discovered that this witch, lawfully executed in 1431, was born in the self-same Lorraine which had been partitioned by the Burgundians in the fifteenth century, and was now partitioned by the Germans once again. Her statues proliferated throughout France, and her shrine at Domrémy became a symbolic bastion of virtuous defiance against an unrighteous invader. Works of science fiction showed how the great battle to liberate the lost provinces would be fought at the villages of Coussey and Neufchâteau on the banks of the Meuse, overlooked by Domrémy itself.[81] The Russian General Dragomirov, a leading advocate of the bayonet assault, wrote articles in praise of Joan, and the patriotic writer Péguy, inspiration to de Gaulle and Pétain alike, wrote not one agonised play on the subject, but two.[82]

All this meant that the inevitable future war had to be fought with the single aim of regaining the lost provinces at the quickstep. No French war plan could be contemplated without an immediate offensive towards the East, and the Germans could absolutely depend upon this taking place.[83] At a lower level of policy, also, the army's embattled military traditions and political tendencies made it highly sensitive to anything that might seem to reflect badly upon its honour or *panache*. It defeated the 'masonic plot' to replace its red and blue uniforms with dingy camouflaged combat dress.[84] It abhorred the republican government's political interference with appointments, and the use of the army to disendow the Jesuits. Still closer to home, it resented the dilution of its long-service, well-indoctrinated conscripts by shorter-term recruits who would be less capable of complex manoeuvres.[85] Most importantly of all, however, it felt that racial, psychological and spiritual factors all made it inappropriate for French soldiers to stand on the defensive.

When he was killed in 1870, Ardant du Picq had already distilled many insights into combat psychology from his own experiences and from the existing rich fund of French military writing. After the *débâcle*, however, the duty of carrying forward his important work fell to a new generation. At first there was understandable turmoil and confusion, as a new army was gradually pieced together, but by the 1890s the newly-established General Staff had become an effective organiser of combat analysis. A Higher War School, later joined by a Centre of Higher Studies, had been set up alongside the historical section of the Staff.

The work of these organisations, although brusquely terminated in 1914, included an unsurpassed scholarly contribution to our understanding of Napoleonic and Franco-Prussian War history. This scholarship was never intended to be impartial, however, since the aim was always to use the lessons of the past as a guide to action in the future. Its authors looked forward not to retirement in a medieval university, as did an Oman or a Delbrück, but to glorious service on the field of honour. Many of the best of them were killed there, including Colin and de Grandmaison – but others found promotion to the very highest levels of command. De Maud'huy went on to command an army, Foch became *generalissimo* of the entire Western front . . . and Pétain eventually became the head of state.

We have already examined Colin's perception that infantry combat was based essentially on firepower. He had his disciples, including Captain 'Danrit', Bressonet, Pétain himself, and several students of the Boer War.[86] In 1904 the official tactical regulations also distanced themselves a little from the emphasis on the offensive that had been evident in the 1884–7 manuals, and still more so in those of 1894–5.[87] The 1905 two-year service law and Japanese successes in Manchuria, however, soon swung the fashion back again. Many French officers thereafter agitated for the regulations to be rewritten in a more 'offensive' manner, and their efforts were crowned with success by the end of 1913, especially in the *Réglement de Manoeuvre d'Infanterie* of 20 April 1914.[88] Ironically, this document starts by stressing the need to avoid constant changes of doctrine, since they would create a *malaise*, yet it represents precisely such a change itself.[89] If we wish to find a reason for French failures in August, the fact that their tactics were only four months old may not be the least relevant.

The 1914 *Réglement* echoed many contemporary opinions when it claimed that

The bayonet is the supreme weapon of the infantryman. She plays the decisive role in the assault, towards which all attacks should resolutely aim, and which alone allows an adversary to be put definitively out of action.[90]

Nevertheless, the whole system envisaged by these regulations is based on an assumption that the young conscripts will be simultaneously enthusiastic for combat, yet calmly obedient to orders. The officers will have been carefully trained both to 'read the battle' and to keep a strong control of their troops. They will not attack until the enemy has first been reconnoitred and suppressed by fire, a process that will be helped by the battalion's two machine guns advancing close behind the skirmish line.[91] As far as it goes, this is scarcely a stupid recipe for victory, although its starting assumptions about the quality and training of the participants were doubtless wildly over-optimistic.

The French problem was not so much how to design theoretically effective assault tactics, but how to train the new two-year conscripts to carry them out reliably. One of the most interesting views on how they should be motivated appeared in de Maud'huy's book *Infanterie*, which was published in 1911. De Maud'huy was a keen advocate of the work of Ardant du Picq, but he had also made a considerable study of the infant science of psychology. He corresponded with Gustave le Bon, among others, whose *Psychologie des Foules* (1895) was a runaway best seller, not least in proto-fascist and military circles.[92] Le Bon was interested in the way large groups were motivated by irrational subconscious impulses, whereas individuals or small groups might behave perfectly sensibly. An individual leader might therefore be able to decide a correct policy, but he would have to use well-chosen symbolism and rhetoric, rather than logical reasoning, if he were to move the masses. By the same token, democratic or parliamentary decision-making was intrinsically flawed, since it would rapidly descend to the lowest, least intellectual, common denominator.

In terms of military tactics, this analysis led to an astonishingly 'modern' understanding of the fears and stresses which the soldier encounters in combat. Like S. L. A. Marshall a half-century later, de Maud'huy had a vision of the individual sensing himself alone in battle, prey to surges of irrational terror because of what he did not see, more than because of what he did. He would fight best if he knew his comrades were watching him, whereas fatigue and enforced immobility under fire would be vital contributors to increased stress.[93] In a passage unconsciously echoing Jackson's view of his 1815 militia at New Orleans, furthermore, de Maud'huy said that normal crowds were incapable of manoeuvres in open terrain, but could fight only from behind barricades or other fortifications.[94] It was therefore the twofold job of the officer to manipulate the military crowd so as to draw the best from it, at the same time as converting it into a higher, more rational and more regular type of organisation. In combat the unit had to be prevented from degenerating into a formless mob for as long as possible, while the enemy had to be encouraged to take on crowd characteristics and abandon his own regularity. Racial factors would be very important in the way these military crowds behaved, since German crowds were less mercurial or audacious than French ones, so officers should have a deep understanding of this element.[95]

The French 1914 doctrines of the offensive drew heavily upon pseudo-scientific social, racial and psychological concepts, but they were mightily reinforced by analysis of the operations in 1870. This was not so much a matter of the most minor levels of tactics, however, where some new assault techniques had been glimpsed during the war, but at a higher level of army manoeuvres. Despite some effective local counter-attacks, the general French posture in 1870 had been passive and defensive, with both Bazaine in Metz and Napoleon III in Sedan contriving to get themselves encircled. Around 1900, therefore, the perception was that defeat in 1870 had been caused by allowing the Germans to seize the initiative and complete their encirclements undisturbed. In a future war this should not be allowed to happen again: it should be the French who seized the initiative, disrupted the enemy's plans, and threw him into paralysis.

This view was expressed most pithily in the work of Colonel (later General) Louis de Grandmaison, a staff officer who had made a close study of recent military history. In 1906 he seemed to lead the widespread reaction to the 'cautious' regulations of 1904, by publishing *Dressage de l'Infanterie en Vue du Combat Offensif*.[96] This was certainly concerned with minor tactics but, despite rhetoric condemning a passive approach, it actually deepened the 1904 call for increased supporting fire in the attacks. As de Maud'huy was to do more thoroughly five years later, he also recognised the growing importance of stress and fear on the modern 'empty' battlefield, and cited some of Le Bon's ideas about crowd psychology.[97] He then proposed that stress should be overcome by solid training, clear doctrine, small group cohesion, and plenty of artillery support. This is far from a pernicious approach to the problem, although in the event it was acclaimed more for its continued acceptance of the frontal attack than for its more useful mechanisms to make such an attack actually work.

More influential within the army – and correspondingly more

damaging to de Grandmaison's posthumous reputation – were to be two staff lectures advocating the offensive which he published in 1911.[98] To these have been attributed the predominance of 'Young Turks' on the General Staff in 1914, and the idea that a 'cult' of the offensive was the touchstone of any commander's efficiency. Two sets of regulations for large unit operations were inspired by de Grandmaison in 1913; although contrary to the popular misconception, his mature doctrines were concerned with minor tactics much less than with 'the engagement of large formations'. They were intended as a contribution to a complex technical debate that had been raging for many years, concerning manoeuvres at the 'operational' level of action.

Everyone agreed that Napoleon had been right to advance into contact with the enemy by using a dispersed fan of army corps, closing up for decisive combat once the key point had been identified – by 1911 no French commander worth his salt would contemplate setting about things in any other way – but the problem was to decide just precisely how Napoleon had intended it should happen, and how he would have applied it himself if he had been born a century later. Since 1895 the accepted orthodoxy had been that each large formation should be preceded by a vanguard, to find the enemy and guarantee 'security-with-confidence' (*sûreté*) for the commander while he stopped to design an appropriate plan. The need for security applied still more to the flanks, where there was an enormous nervousness that the Germans would apply their customary doctrine of encirclement. Yet in their exercises the French found that reconnoitring vanguards tended to fight only withdrawal actions when they met the enemy, while fear for the flanks tended to suck the follow-up troops outwards rather than forwards, thereby dissipating their offensive power.[99] The overall effect was one of hesitation, and the initiative was left with the enemy.

In de Grandmaison's two lectures, therefore, he argued that the French fan of dispersed formations should not be allowed to open out when the Germans were found, but should redouble its forward march. The attacking troops should also be well concentrated towards the front, not echeloned rearwards in an over-complex succession of detachments and reserves. Apart from anything else, he pointed out that such an arrangement would possess a simplicity that might, at long last, allow every participant to understand what was expected from him. De Grandmaison's message was that the Germans' strict march procedures were vulnerable to dislocation if their preconceived timetable could be upset. They themselves were aware of this vulnerability, however, and hoped to keep the French at bay through the use of strong and aggressive forces in their vanguards. By adopting a more flexible and tentative approach, with weaker vanguards, the French were therefore voluntarily abandoning the initiative to their opponents. They would do better to take a leaf out of the Prussian manual and batter down the initial opposition while it was still weak. By 'biting fast and deep'[100] the French could upset the rigid enemy plan and win an important advantage in both morale and operational posture. They could win security not by a physical defensive screen, but by a dynamic offensive movement.

In the event, in 1914, the Germans won some important advantages precisely because their vanguard screens in Belgium were heavier and more aggressive than those of the French. Both sides also suffered from just the type of worry about flanks that de Grandmaison had warned against,[101] and it was often by threatening encirclement that the Germans were able to maintain their mobility forward into northern France. They seized the initiative in very much the manner that had been expected, and de Grandmaison cannot be faulted for these aspects of his analysis.[102] Where his recipe came unstuck, however, was in the way his brusque attacks were actually executed.

In Joffre's conception of Plan XVII, as it had evolved by the middle of August, there were to be three separate offensives – an early right hook from Nancy passing Metz to the south, a left hook attacking frontally through Charleroi to repulse the enemy spearhead in Belgium, and a slightly delayed main assault through the Ardennes to cut through the Germans' rear.[103] Each attack consisted of two armies in line abreast, with those in Lorraine and Charleroi pressing forward for a frontal clash in very much the concentrated formation that de Grandmaison would have wished. The attack into the Ardennes, however, was more tentative, more widely dispersed and fragmented. It cannot be attributed to de Grandmaison's vision, but to Joffre's more cautious and traditional idea of waiting to collect intelligence, and then launching a central breakthrough 'in the style of Austerlitz' where the enemy was believed to be weak. When the intelligence turned out to be wrong, and the enemy was found to be advancing through the Ardennes with numerical superiority, both the strategic and the operational conceptions collapsed. Equally, Lanrezac's Fifth Army at Charleroi lost its offensive impetus when it found an attractive position along the south bank of the Sambre. Far from making an *attaque à outrance*, Lanrezac stuck fast and fought defensively for three days before following the BEF in the general rearward movement.[104]

The best test of de Grandmaison's theory of the offensive should perhaps be sought in the Lorraine attack conducted by Dubail's First and de Castelnau's Second armies. Between 19 and 20 August these came into action at Sarrebourg and Morhange respectively, on an overall frontage of around thirty miles. They attacked, but often without following de Grandmaison's principles at all closely. For a number of reasons they were held fast by an enemy standing in position, and then retreated due to counter-attacks and threats to their flanks: it all made rather an ignominious end to three years of doctrinal theorising and agitation.

At Sarrebourg the French VIII Corps attacked against approximately its own numbers on a seven-mile front to the north-east of the town. An initial frontal attack by three regiments from the town itself was stopped at the foot of a long open slope by artillery and machine gun fire. The assault made better progress when it was renewed on the 20th with increased artillery support, but an enemy counter-attack finally drove it back. In the afternoon the French had to fight a defensive house-to-house combat in Sarrebourg, to cover their retreat.

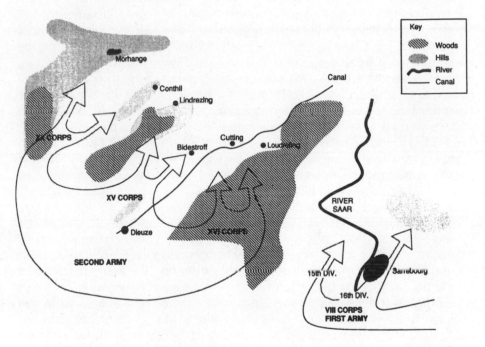

The battles for Lorraine, 19–20 August 1914.

Further to the left the terrain was more broken, wooded and undulating. Despite a failure to assault at dawn, two regiments of the 15th Division each took their initial objectives – one by skirmisher infiltrations, the other by cold steel alone. Heavy artillery fire and a solidly fortified infantry position soon halted both attacks, however, and a third regiment failed to take even its first objectives. Pressure on the flanks forced a retreat by lunchtime, although the French observed that their own artillery could be just as effective in the defence as the Germans'. The 'decisive battle of Sarrebourg' was thus a very short-lived and small-scale episode indeed, with little more than 24,000 French troops engaged for little more than a day, and around 8,000 casualties. This was the highest loss of any corps in First Army, but by nineteenth-century standards it was scarcely a shocking figure for such a crucial engagement.[105]

A very similar story, albeit on a rather larger scale, can be told of the battle of Morhange. Second Army attacked with three corps in line on a twenty-mile front, against an essentially equal enemy force. On 19 August Foch's XX Corps on the left established itself on the ridge running east towards Morhange village, but there was some heavy fighting. De Grandmaison himself was wounded twice, and his regiment was to lose 1,100 casualties in 24 hours. Meanwhile, in the natural amphitheatre below, an eyewitness reported the 37th regiment's '*magnifique*' advance under shelling, in bounds by half platoons (*sections*), to capture Conthil. Alas, they were evicted at 7 o'clock next morning by a counter-attack, although pursuit was halted by two companies and two machine guns from the 26th regiment, supported by artillery. The 79th regiment subsequently fought a heavy battle around Lindrezing, but neither side seemed to have arranged proper artillery co-ordination, and there was an echeloned disengagement in the early afternoon. Foch's intended grand offensive to capture Morhange on the 20th had been both pre-empted by the enemy's own offensive, and countermanded by the army commander's orders. It was scarcely what de Grandmaison can have had in mind in 1911.[106]

On the centre and right of de Castelnau's army the advance was no more spectacular. XV and XVI Corps made a few gains, notably the village of Bidestroff, but were unable to debouch either from there or from the edge of a long wood facing the fortified villages of Cutting and Loudrefing. When the advance bogged down on 19 August there was a good deal of digging in, with some units 'attacking' only by fire.[107] Then next day the German counter-attacks came in as heavily as at Conthil and Lindrezing, making gains particularly along the junction between the XX and XV Corps areas. Bidestroff fell after severe shelling and two assaults, and by mid-morning de Castelnau was facing the possibility of a central breakthrough of his position without reserves to fill the gap. He became convinced that he had to pull back his entire army out of contact.[108]

One important feature of the French system was that above the level of the three-battalion regiment, each unit had two sub-units. There were two regiments to a brigade, two brigades to a division and two divisions to a corps. This meant it was awkward for a commander to keep a proper balance between the troops committed to action and those kept in reserve, since an attack 'one up' would leave a massive 50% in reserve, while 'two up' would leave nothing at all. At Sarrebourg-Morhange the two armies certainly attacked in a long line of divisions, usually without a significant operational reserve; and this surely deprived them of a measure of resilience. De Grandmaison had happily accepted this effect in principle, but only on condition that the line was not too long and all its parts attacked simultaneously and vigorously. In practice, however, his preconditions were not properly met. The attack frontage of First and Second armies in 1914 was about twice as long as he would have envisaged, and the advance of each element was not properly synchronised. This in turn opened the door to all the anxieties about encirclement that he had originally feared and tried to avoid.

In summary, there were relatively few regiments involved in the French Lorraine offensive that did not sooner or later make an attack of some kind, but in few cases were these attacks run according to the textbooks. Many were only partial, or poorly supported by artillery, or awkwardly thrown on to the defensive by the enemy's unexpected advance. At the level of corps and army, furthermore, the higher commanders do not always seem to have wished to push forward *au fond* in the way de Grandmaison might have wanted.

Some French attacks succeeded through luck, surprise, good artillery preparation or the skilled use of phased bounds by skirmishers in open order; and rather more German attacks did so. It was certainly not the case that every attack was inevitably doomed to collapse. However, there is no doubt that the attack had become more difficult than in the past, especially if the enemy was alerted and enjoyed an open field of fire. When the conditions for an attack were not especially favourable, therefore, it might all too easily resemble the following action that took place soon after the battle of the Marne:

> . . . as soon as the first silhouettes are outlined against the bare plateau, they are pinned to the ground by the fire of infantry and artillery. No cover, no fold in the ground permits an effective movement. The losses are high for an advance of twenty metres. Give oneself help by one's own fire? So be it! – the section leader commands "Fire at Will" . . . but against what? The target remains invisible and our men give an impression of firing straight ahead of them, without looking. Soon there is immobility, for anything that moves is spotted. Face down to the earth one waits for the storm to pass, and any movement of retreat itself becomes difficult before nightfall.[109]

Just as many American Civil War attacks failed when organised in defiance of Hardee's system of tactics, so there were many French mistakes in 1914 made despite, rather than because of, de Grandmaison's recommendations for the offensive. It would take experience, time and painful losses to learn how to put the essence of his vision to practical effect. On the other hand, it is probably quite fair to blame him, along with most other designers of tactical methods, for imposing unrealistically complex ideas upon ordinary soldiers. In common with the rest of his generation, furthermore, he may be blamed for leaving the army badly prepared for defence.

A striking feature of all these battles is that the Germans could apparently bring something extra to bear in their attacks – probably a greater weight of supporting artillery[110] – which for various reasons the French seemed badly equipped to counteract. Perhaps French trenches were less deep, and French soldiers less well indoctrinated in the idea that the defensive might sometimes be honourable. Maybe there was even some truth in the often-claimed racial differences between the stolid Teuton and the impatient Latin. Whatever the true reason, there was scarcely anything radically new in the general shape of the Lorraine battles of 19–20 August. The frontages were longer than might have been expected in Napoleonic times, but the duration was, if anything, shorter, and the leaders less experienced in the art of battle-handling.

4

1916–1945: The Alleged Triumph of Armour Over Infantry

If our ideas about the First World War are tinged with a lack of sympathy for pre-war tactical analysts, they are also marked by a correspondingly strong admiration for the post-war champions of tanks. We tend to think of the war as a demonstration that infantry had become ineffective, while armour could break through. We see it as a victory of technology over outdated spiritual principles.

There is a lot more to the story than this, however, since the tank turned out to be considerably less useful than one might imagine, and the infantry continued to enjoy a very great importance on the battlefield. Its evolution did not stop in 1914, but continued in the same directions it had been taking since Napoleonic times. The skirmish line became ever looser, while individual skill and initiative became ever more important.

Whereas Wellington's two-deep line had been considered a dangerously loose formation in 1808, the Japanese of 1904 were finding that even a solid single line was too heavy. By 1914 the accepted arrangement had become a line with intervals of several paces between each man. It is in this formation that we tend to visualise First World War soldiers 'going over the top' in the attack, or firing over their trench parapet in the defence. If we stopped to think about it a little further, we might add that each attack would probably be made up of several battalions or at least several companies at a time, and that each trench would be essentially linear. This is the common image we have of what fighting was like between 1914 and 1918.

In fact, however, tactics underwent considerable modifications as the war progressed. The formations and methods used by the end were very different indeed from those which had been used in the first two or three years. At some time after the Battle of the Somme (1916), there took place yet another lightening of the battle lines in almost every army which was engaged. Instead of attacks by whole battalions or companies, the new tactics called for skirmishes, infiltrations or raids by platoons or even squads. As for defensive positions, they were often given a strength of only one machine-gun team. Continuous lines of trenches started to look decidedly antiquated. By 1918 the infantry had increased its dispersion to an extent which would have been unimaginable even four years earlier. There had been nothing less than a complete and rapid reorganisation of tactics to meet the various new challenges of the battlefield.

The story of the development of infantry tactics in the First World War is really the story of what the Germans called the 'storm trooper'. He was to be a foot soldier of a new type, adapted for the conditions of the modern battlefield and employing a range of new weapons to the full. He would be given even more flexibility and independence than ever, and would fight in particularly small groups. In the attack he would skirmish forward cautiously and attempt to infiltrate through an enemy position, rather than tackle it head-on. In defence he would act as an immediate counter-attack element in support of a dispersed 'web' of small strongpoints. The enemy would be enmeshed in the web and then struck with violent blows at the moment when he was least prepared to receive them.

The above is a summary of modern battle as it was generally accepted by the end of 1918. It was a conception which had apparently first occurred to a French officer, Captain André Laffargue, when his attack was held up on Vimy Ridge on 9 May 1915.[1] He realised that he would have done better had he sent forward small groups of men to engage the enemy from the flank, as well as some light mortars to provide indirect fire at the time and place it was required. It was apparent to him that the trust placed in formal artillery support was exaggerated, since the artillery of 1915 lacked the flexibility to respond immediately to the infantry's need. Laffargue's vision was therefore of an infantry unit which could fight its own way forward with its own weapons.

In 1914 the infantry had been armed with three different weapons – rifle and bayonet, pistol for officers, and perhaps the occasional machine-gun. After 1918, by contrast, Colonel de Gaulle reported that the French infantry needed a total of sixteen types.[2] Many new weapons had emerged in the course of the war; a range of mortars, a range of light machine-guns, a flamethrower and, perhaps most significantly of all, a variety of hand grenades. It was in the First World War that the hand grenade displaced the bayonet as the arm par excellence of the close quarter fighter. It was the principal weapon of the storm trooper.

A squad of storm troopers could be built around a mortar, a light machine-gun or both. These weapons would provide heavy covering fire while the riflemen and bombers worked their way forward in short rushes to the enemy's line. While they were still keeping their heads down from the covering fire, the enemy soldiers could be bombed out of their trenches.

Ironically it was not to be the French army which first put these ideas into practice but the German. In the First World War as in the Second it was the Germans who felt the greatest need for innovative fighting methods, since they suffered from a marked numerical and material inferiority relative to their enemies. For every shell they fired, half a dozen or more seemed to be returned. In these circumstances it became especially important for them to use high quality soldiers and effective tactics. They found they simply could not afford the same margins of error which the Western allies seemed to enjoy.

This view is faithfully reflected in the memoirs of Ernst Jünger, a German subaltern who received storm troop training. By the end of the war he had been wounded more than fourteen times, and yet he remained a keen enthusiast for trench fighting. Despite the 'physical effects on his memory of repeated blows from great pieces of metal,'[3] he has left us a cogent picture of these developments in German tactics.

It was in the Battle of the Somme (1916) that he first realised that the material balance had decisively shifted. The terrific barrages which he underwent . . .

. . . first made me aware of the overwhelming effects of the war of material. We had to adapt ourselves to an entirely new phase of war. The communications between the troops and the staff, between the artillery and the liaison officers, were utterly crippled by the terrific fire. Despatch carriers failed to get through the hail of metal, and telephone wires were no sooner laid than they were shot into pieces. Even light-signalling was put out of action by the clouds of smoke and dust that hung over the field of battle. There was a zone a kilometre behind the line where explosives held absolute sway.[4]

All this tended to isolate the front line soldier and thow him back on his own resources, or at best, those of his NCOs:

One hears it said very often and very mistakenly that the infantry battle had degenerated to an uninteresting butchery. On the contrary, today more than ever it is the individual that counts.[5]

This, indeed, was a realisation which had been growing for many decades among German analysts; but it finally came to fruition only after the Battle of the Somme. Before then there had still been a counter-current of thinking which had clung to the old idea of massed or at least linear action. On the Somme, however, the shape of the future was finally appreciated by the Germans in all its aspects:

Chivalry here took a final farewell. It had to yield to the heightened intensity of the war, just as all fine feeling has to yield when machinery gets the upper hand. The Europe of today appeared here for the first time on the field of battle . . . .

The terrible losses, out of all proportion to the breadth of front attacked, were principally due to the old Prussian obstinacy with which the tactics of the line were pursued to their logical conclusion.

One battalion after another was crowded up into a front line already over-manned, and in a few hours pounded to bits.

It was a long while before the folly of contesting worthless strips of ground was recognised. It was finally given up and the principles of a mobile defence adopted. The last development of this was the elastic distribution of the defence in zones.[6]

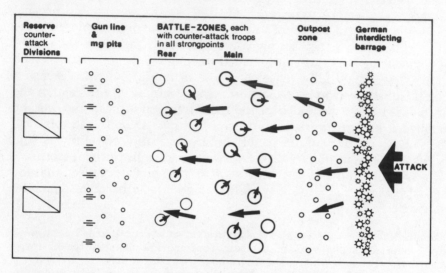

*German 'web' defence, First World War, after the Battle of the
Somme. Attacks are split up and canalized as they advance into
the web, and are then counter-attacked.*

After the Somme the Germans abandoned their continuous lines and moved to a system of defence designed to save casualties. This was the 'web' defence, with a front line consisting of scattered outposts covering successive deep layers of a main position. Behind these, again, would be further scattered machine-gun nests backed up by strong reserve forces. In every layer of the defence there would be groups of specially-trained storm troops, poised to counter-attack any enemy lodgement.

In 1937 Rommel described the system as follows:

Modern defence organisation differs greatly from 1914. Then we had a front line with the remaining troops disposed in a second line. Today a battalion position consists of an outpost line and a main battle position through which the forces are organised in great depth. In an area 1100–2200 yards wide and deep, we have dozens of mutually supporting strongpoints garrisoned by riflemen, machine-guns, mortars and anti-tank weapons. These dispositions cause the enemy to divert his fire and the defence to concentrate on its own fires. Local manoeuvre under covering fires is possible, and aggressive counter-measures can be instituted should the enemy succeed in penetrating

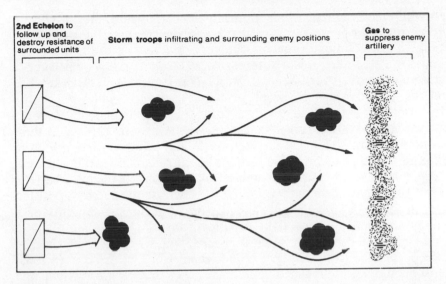

German offensive technique, 1917–18.

the main battle position. The enemy has a long and very difficult road before breaking through.[7]

If we make allowances for the increased emphasis on anti-tank guns by 1937, we find that Rommel here describes exactly the layout adopted for German defensive positions in 1917 and 1918. It was a system which depended upon great depth; upon the use of small groups delivering flanking rather than frontal fire with their machine-guns; and also upon a decentralisation of initiative to local counter-attack forces. Another aspect which Rommel does not mention, but which was equally important, was the arrangement for gaining artillery superiority. Because the Germans expected to face superior enemy artillery strength, they tried to nullify its effects by the use of ground. Their forward outposts would be deployed on the crest of a hill so that as soon as enemy infantry had advanced through them into the web, its supporting artillery would be denied observation of the battlefield. The Germans alone would be in a position to bring observed artillery fire to bear at the point of contact on the reverse slope. The enemy spearheads would thus be isolated from their supports, confused and surrounded by hostile strongpoints.[8]

A further refinement was added to this system in 1917, when the Germans sometimes took a step back with their front lines just as the attackers commenced their barrage. The heaviest and best-prepared fire would thus fall into a vacuum, and would serve only to make the ground more difficult for the attacking troops to cross. The Germans would meanwhile have retired to new positions in rear, where they would be waiting unscathed for the arrival of an exhausted and disorganised pursuit. This technique proved to be very effective on the Chemin des Dames in 1917, and in the summer of 1918 it had almost become a routine procedure. In the Second World War it was again often used to good effect on the Eastern Front.

It is important to realise how far from traditional conceptions these German defensive tactics had moved, if we are also to appreciate their new methods of attack. What seemed to happen for their offensive technique was that many of the features of the defensive battlefield were almost 'turned inside out' and superimposed upon the enemy's line.

Thus the aim of the attack would be to penetrate the enemy's position in depth at the first onslaught, and to surround his strongpoints with small groups built around light machine-guns and mortars. Initiative would be fully decentralised to these groups, whose only orders would be to keep pushing forward. A second echelon of more conventional infantry could then follow up to eliminate the enemy positions which had been cut off. In order to achieve artillery superiority there would be only a very short barrage; but one of great violence and depth. It would play up and down the enemy's area from front to rear and then back again, and would particularly aim to neutralise his artillery with a surprise gas bombardment. It would also be followed up very rapidly by the storm troopers who might hope to reach as far as the enemy gun line in their first bound.

In the attack as in the defence these tactics were designed to spring a series of nasty surprises on the enemy, to deprive him of his artillery support and to surround bodies of his troops so that they could be wiped out in detail. They represented a coherent way of looking at the whole battlefield, and in practice they turned out to be very strong indeed.

It is General Hutier who is generally credited with making the transformation from the Germans' defensive system into the offensive one described above, following his experiences at Riga on the Eastern Front. In the West his ideas were first applied on a grand scale during the Battle of Cambrai in late 1917. This battle had started with an astonishing British surprise attack using tanks en masse, which had been stopped only on the Germans' final defensive line. The 'Land Ironclads' had swept up a great number of prisoners and trophies in their initial onslaught, although it must be admitted that their many defects had shown up clearly to those who had eyes to see. To most people it seemed that an answer had at last been found to the supposedly impenetrable German defensive system.

It was in a large scale counter-attack that the Germans then gave their first great demonstration of the new offensive tactics, winning back lost ground almost as quickly as the tanks had taken it. Without the benefit of any revolutionary new weapon, they too showed the potential power of deep infiltration and surprise. It is ironic that the enthusiasts of armour were seeking to exploit precisely these qualities in their own way, since the Germans showed they had nothing to learn in those areas. At Cambrai the combination of storm troopers and well-orchestrated artillery proved to be as much of an answer to the problems of modern battle as the tank itself.

Ernst Jünger took part in the Cambrai counter-attack and concluded that it was the courage and daring of well-formed infantry which held the key to tactical success. Even in the summer of 1918 he remained unimpressed by tanks, and experienced a sense of sadness mixed with a slight awe at the hopeless plight of the men who went to war in them. When he inspected a number of these machines which had been destroyed by German shells he found:

> . . . they were all in a pitiable plight. The little cabin of armoured plate, now shot to pieces, with its maze of pipes, rods, and wires, must have been an extremely uncomfortable crib during an attack, when the monsters, hoping to baffle the aim of our guns, took a tortuous course over the battlefield like gigantic helpless cockchafers. I thought more than once of the men in these fiery furnaces.[9]

For Jünger it was artillery and machine-guns which represented the technological side of modern warfare, not the tank. For him, as for the tactical analysts of the nineteenth century, the answer to more difficult battlefield conditions seemed to lie in improving the individual qualities of the fighting man himself. The more each man was cut off from his neighbour by increasingly dispersed formations, the more he would need a high standard of morale and training. This was not a reaction based upon pure theory or 'military spiritualism'; it was an empirical finding derived from hard experience. Jünger was quite specific about what he thought should be done to foster the required qualities in the soldiers:

> Above all, I devoted my efforts to the training of a shock troop, as it became more and more clear to me in the course of the war that all success springs from individual action, while the mass of the troops give impetus and weight of fire. Better command a resolute section than a wavering company.[10]

This view gained ready acceptance in many other quarters, for by the end of 1918 the new infantry tactics had been spectacularly demonstrated once again by the Germans in their spring offensives, and to some extent copied by the allied armies after that. The allies had taken longer than the Germans to reform their methods; but by the time of their late summer attacks they had little still to learn. They were also able to incorporate large numbers of tanks into their offensives, which in fact strengthened them considerably for as long as the tanks did not break down.

101

After the war the British fully recognised the imperative need for greater initiative and independence at the level of the private soldier and his NCO. In 1932 an official War Office pamphlet was stating that training should be designed to encourage 'Our riflemen and light automatic gunners to be formidable fighting men, fit, active, inquisitive, and offensive, confident of making ground with their own weapons.'[11] This was precisely, and consciously, a reflection of what the German storm troopers had been trying to achieve.

Another example of a similar reaction came from Tom Wintringham, who for a short time led the British Battalion in the Spanish Civil War. He felt that the value of tanks in 1939 had not improved since 1918. They could sometimes be a useful auxiliary, but could not achieve decisive results:

The exponents of mechanisation say that the tank is the solution of the problem of how to make attack easy and profitable. The tank is no such thing. It is a weapon of opportunity to be used rarely and by keen-eyed men who know its limitations. It is short-sighted, noisy, necessarily nervous even if manned by brave men, and a bad gun-platform when moving. When stationary it is a good target for artillery, and if within reach of opposing infantry it can be stalked . . . .[12]

The verdict of Spain – which I believe will also be the verdict of the next great war – is not that the tank is an utter failure, but that it has restricted use as a weapon of opportunity. Tanks cannot possibly – whatever size you make them, or however many you make – replace infantry as the basis of the army, because they cannot hide as infantry do, go to ground, become dangerous vermin hard to brush out of the seams of the soil.[13]

Wintringham concluded, as Jünger had before him, that it was still infantry infiltration tactics under NCO's of initiative which held the key to offensive warfare. In the defence an 'elastic' line with strong local counter-attack elements was required. Both Wintringham and Jünger had seen the face of modern war at close hand, and to both it was the training and morale of the individual fighting man which appeared to be of paramount importance. Tanks and aircraft held few terrors for either of them.

In France it was paradoxically an enthusiast of armoured troops, Colonel Charles de Gaulle, who pointed out the need for improved moral qualities in the circumstances of modern war:

Even in the gloomy hecatombs to which the exclusive system of the nation-in-arms led during the Great War, the superiority of good troops was abundantly clear. How else is one to explain the prolonged success of the German armies against so many different opponents? . . . Witnesses of the final engagements have not forgotten those proud picked troops, broken to every test, who in every action led the 'main effort'[14] . . . or again; . . . as the danger and desperation of fighters on the battlefield increases, moral cohesion becomes more and more important.[15]

De Gaulle's vision of modern war was a fairly close echo of Ardant du Picq's. In both cases there was a realisation that battle was becoming more difficult and dangerous, and that this imposed unbearable strains upon troops who lacked careful training or the 'military spirit'. A small élite force would therefore be preferable to a mass of reluctant conscripts. It would alone remain manoeuvrable, aggressive and able to use modern weapons to good effect. Technology certainly had a place in this picture; but it was man himself who provided the most important element.

In the event it was found that neither man nor the machine was capable of achieving a break-through on the Western Front. On many occasions a successful 'break-in' took place, led either by tanks or by stormtroops; but the principal difficulty was always one of exploitation. Time after time it was found to be impossible to press forward once the initial bound had lost momentum. Just as in 1877 the first line of trenches at Plevna had usually fallen to the Russian attacks, so in the First World War it was almost always possible for an attack to make headway through the forward parts of an enemy position. In both cases, however, it was very difficult to renew the impulsion after the original success had run down.

The transition from break-in to break-through is a ticklish problem at the best of times, and even the highly manoeuvrable armies of Napoleon had achieved it but rarely. In the conditions of the Western Front it was made very much more difficult by a combination of the ground, the weapons, the poor signals equipment and the transport facilities available. Taken together, these factors meant that it always took an attacker longer to fight through an enemy defence zone than it took the enemy to bring up significant reinforcements:

This problem is, broadly, how to maintain a continuous and sustained effort throughout an attack, so as to take instant advantage of initial success and to achieve complete victory before the enemy reserves can intervene. It really resolves itself into a time problem between the onrush of the attacker and the enemy reserves.

In the late war the race nearly always went to the defender.[16]

Movement over the battlefield was hampered by mud, shell holes and the heavy loads which each man carried. It could be seriously delayed by relatively small pockets of resistance or by interdicting artillery barrages. It lost cohesion as the men took cover and dispersed. Without efficient signals it was difficult to draw units back together for further operations, or even to explain the position on the ground to higher headquarters. Orders could not be issued in time, or with an adequate basis of information to make them sensible. In these circumstances the progress of any attack was bound to be slow and hesitant, even if the enemy himself had been successfully overcome.

On the other side of the coin lay the conditions prevailing beyond the battlefield. Here there were good telegraphic communications and good railway lines to bring up reinforcements very rapidly indeed. While an attacker was blundering around within a wilderness of churned-up mud, the defender could concentrate fresh formations at his leisure; deploy them behind the battle zone in a new line; and await the outcome with confidence. The real disparity in the First World War was not therefore between the attacking infantry in the open and the defending machine-gunners in cover; but between the attacking mob on foot and the defending reserves steaming up in well organised railway trains.

In these circumstances the theoretical importance of the tank lay in its ability to move men, weapons, supplies and radios quickly over a broken battlefield. It seemed to offer an attacker the chance to win the race to the defender's railheads. In practice, however, this conception consigned the tank to too much of a 'transport' role for the tankers' liking, and too little of a 'cavalry' role. Experiments with armoured command vehicles, armoured personnel- or ammunition-carriers and armoured gun tractors never had quite the same visionary fire behind them as did the use of tanks for an assault role. By the end of the war, admittedly, tank officers such as J. F. C. Fuller were thinking in terms of deep penetration with 'cruiser' tanks rather than simply short assaults in close support of infantry. Despite copious lip service, however, not even Fuller's famous 'Plan 1919' seems to have made an adequate provision for armoured carriers. It was the 'shock' effect of massed armour, and the psychological disruption it could cause, which was given more attention than the mundane task of passing a force of all arms safely and quickly through the combat area.

To some extent this emphasis on assault was a matter of technology. The early tanks were heavy and slow, designed for infantry support and liable to break down almost as often as they worked. To many observers it did not look as though very much could be demanded of them further afield than the enemy's first line of trenches. At Cambrai under 40% of tank losses had been caused by enemy fire: the rest had been breakdowns or ditchings.

Another reason for the stress on assault, however, was the bitter institutional battle which the tankers had to fight within their own armies. The infantry were their natural allies in these arguments, against the cavalry who initially felt their role was threatened. It was therefore natural that the thoughts of the tank enthusiasts first moved in the direction of infantry support, and then progressed to ambitious claims that tanks could replace the cavalry. Too often the latter led them into rhetorical extremes and made them imagine that the tank could do many things which in fact it could not.

It was a matter of profound disappointment to the British Tank Corps that no genuine break-through was ever achieved on the Western Front by armour, and yet in Palestine Allenby successfully achieved one with cavalry at the Battle of Megiddo (1918). Indeed, the pattern of Allenby's operation was full of lessons for the future, since he aimed at a truly deep penetration with close air co-operation. He was bombing enemy headquarters and supply dumps in a way that Fuller would have heartily approved. His success, however, served to strengthen the believers in cavalry rather than those who dreamed of an armoured *blitzkrieg*.

When the war was over and the British Army had stood down from its continental role, a strong current of opinion started to flow against the tanks. They were expensive, specialised, less than decisive on the battlefield, and apparently subversive of military spirit. Just as the French Chasseurs had been suspect in the 1830s for their over-enthusiastic officers and under-drilled soldiers, so the Tank Corps in the 1920's appeared to be a very dubious luxury for an army dedicated to imperial policing. Fuller, its standard-bearer, came to be considered a 'military bolshevik' in many circles.

Part of the problem was Fuller's apparent reversal of Napoleon's maxim that in war 'the moral is to the physical as three to one'. For this time-honoured formula he substituted the suggestion that 'tools and weapons, if only the right ones can be discovered, form 99% of victory'.[17] In common with de Gaulle, Fuller favoured an army composed of a small but professional body of experts; but unlike his French counterpart he did not seem to pay much attention to the moral elements in battle at any level lower than that of the commander in chief. Fuller's view that matériel alone was decisive was certainly overstated, and indeed represented a reversal of his own earlier beliefs. Before the war he had been a keen student of Napoleon and of 'Sir John Moore's system of training', while in 1913 he had written that 'to will success is all but equivalent to victory'.[18] For him the World War made a decisive break with the past whereas for many soldiers, including some of the continental exponents of armour, it had served to reinforce a good part of the accepted doctrines of morale.

It is possible that Fuller's excessively sharp reaction to his critics, as well as his penchant for mysticism, did more damage to the cause of armoured warfare in the British Army than all the cavalry messes put together. When he subsequently embraced fascism and struck up friendly relations with Hitler and Guderian, he did nothing to atone for his earlier errors of judgment.

In Guderian, by contrast, the Germans had an officer who, although junior to Fuller and lacking any tank experience in the World War, had a much better sympathy with the military ethos. He was personally able to work well within the army and get the most out of it. For him, moreover, the problems of mechanisation had first been posed as a matter of motor transport rather than of assault. This was necessarily so, since the German Army had been forbidden to build tanks by the Versailles treaty. For Guderian the debate was not at first a conflict with the cavalry, as it had been for Fuller, but with those who regarded transport as essentially something which took place behind the lines. Thus he says that in 1924:

I expressed the hope that as a result of our efforts we were on the way to transforming our motorised units from supply troops into combat troops. My inspector, however, held a contrary opinion, and informed me bluntly: 'To hell with combat! They're supposed to carry flour!' And that was that.[19]

Guderian did not fritter away the 1920s in metaphysical attempts to prove that flannel shirt was inferior to armour plate as a protection against machine-guns. Instead, he read widely in all aspects of motor transport, tank tactics and mobile warfare. Unprejudiced by the debilitating committee warfare of the Western Front, he was able to look at the matter in the round. Nor did he make up his mind about the whole question until as late as 1929, after he had seen how the Swedish army employed its German-designed tanks:

In this year, 1929, I became convinced that tanks working on their own or in conjunction with infantry could never achieve decisive importance. My historical studies, the exercises carried out in England and our own experience with mock-ups had persuaded me that tanks would never be able to produce their full effect until the other weapons on whose support they must inevitably rely were brought up to their standard of speed and cross-country performance. In such a formation of all arms, the tanks must play the primary role, the other weapons being subordinated to the requirements of the armour. It would be wrong to include tanks in infantry divisions: what was needed were armoured divisions which would include all the supporting arms needed to allow the tanks to fight with full effect.[20]

Guderian had grasped the truth that whereas the tank was far from invulnerable in itself, its mobility could also be applied to other arms of the service to make a balanced all-arms force. What was needed was a way of bringing motor transport, in all its forms, into the fighting line. The partial successes gained by tanks at Cambrai or Amiens could then be married up to the partial success of motor transport, as represented by the 'Taxis of the Marne' or the endless crocodile of lorries which had sustained Verdun. The net result would be to restore mobility to the battlefield as a whole.

This line of thinking led to the creation of the German 'Panzer' Division, as an all-arms force which combined the best qualities of infantry, artillery, armour and motor transport. It displayed a more complete integration of these different elements than could be found in any other army, and it was based upon a better appreciation that the tank was no more than *primus inter pares*. The tank was seen as a specialised machine which could help the other arms to accomplish their tasks, just as they in turn could help it. In 1939–40, indeed, the tank still had many technical deficiencies which seriously limited the scope of its action. It was not to be until the appearance of the Russian T34 in late July 1941 that anyone achieved a really good combination of armour, armament, reliability and speed. Until then all tank designs had fallen a long way short of what had been hoped.

Even with inferior tanks, however, the Panzer Divisions proved to be capable of great achievements against enemies who lacked equally flexible formations of all arms. In their great 'blitzkriegs' of 1939–41 the Germans overran country after country with apparent ease. By their agile orchestration of novel weaponry they turned bewilderment into panic, and by the rapidity of their advances they spread terror far and wide in the enemy's rear echelons. In these campaigns they seemed to be unstoppable. Not a few observers started to believe that, after all, the infantryman had finally been superseded by the tank.

In none of the early blitzkriegs, however, did the Germans meet the sort of opposition which might have exposed the tank's real weakness. Enemy strongpoints could generally be avoided and encircled by the use of mobility, rather than assaulted head-on. In these early battles every obstacle or fortified line which the Germans encountered was found to be either full of holes or weakly held. Thus they were able to cross the Russian fortified line on the river Bug, 22 June 1941, before it had been manned. In May 1940 they outflanked the French Maginot line and crossed the Meuse at Sedan. Facing them on this occasion was a second-rate French formation which lacked both tactical depth and moral conviction. Even so the battle was a fairly close-run thing. It was won more by air power, artillery and infantry than by the tanks themselves. As Guderian later admitted in his memoirs, '. . . the success of our attack struck me as almost a miracle'.[21]

When they were not crossing obstacles in extremely risky operations the Germans were often racing across relatively undefended territory to forestall the creation of fresh obstacles further to the rear. Much of the first four or five months of the Russian campaign had this aspect, with the armoured spearhead commanders constantly urging rapid advances to disrupt enemy building of new fortified lines. Far from exhibiting confidence in the assault strength of their tanks, therefore, they were actually doing everything they could to avoid a phase of positional warfare. The real strength of armour, in fact, lay not in battle but in the pre-emption of battle.

Quite apart from fortified lines, the Germans were also anxious to avoid counter-attacks by enemy reserve formations. The essence of 'blitzkrieg' was to break through the enemy's defences before he had time to gather his reserves together. It should disrupt his counter-attacks before they had been formed. Only then would the apparently decisive advantages of the defensive be truly broken.

The early German blitzkriegs certainly succeeded in all this, and were able to brush aside the partial counter-attacks launched by armies which were still thinking in terms of pre-mechanised movement rates. Even so it was with considerable nervousness that the German high command ventured forth from its bridgehead at Sedan in 1940, and on several occasions it almost called off the whole operation for fear of the damage a counter-attack might inflict. When the British did manage to launch a local attack at Arras it caused disproportionate confusion in the German spearhead. In Russia, also, there was considerable disruption of the German plan when Marshal Timoshenko mounted a large counter-attack south of Smolensk in the second half of July 1941. In both these cases the counter-attack had the effect of forcing battle upon formations which were trying to avoid it by the use of mobility. More of the same might well have stopped the Germans in their tracks.

The secret of the 'blitzkrieg' lay in moving a mechanised all-arms force through an enemy's front line before he had time to consolidate it, and then playing havoc in his rear areas before he could mount a counter-attack or build a fresh defensive line. This process did not rely particularly upon tanks – and in Norway it was effectively completed without their assistance. What it *did* require was rapid transport, surprise and an overawed or demoralised enemy.

An alerted enemy, on the other hand, could do several things to make such an attack stop and fight, for without the benefit of surprise a great deal of the power of the offensive melted away. Thus the drive into Russia was finally halted outside Moscow and Leningrad once the onrush of the German spearheads had been delayed sufficiently for defensive lines and counter-attack forces to be organised. The following year the same pattern was repeated at Stalingrad with more spectacular results, while at Kursk in 1943 the Russians were able to build defences in great depth which stopped the attack almost before it had begun.

The creation of effective mobile reserves was another important measure which could stop an armoured spearhead. In the First World War these had always been available in good time by the use of rail transport; but in the Second War the railways had become too vulnerable and inflexible for this role. The battle now moved faster and so mobile reserves now had to move on wheels and tracks, exactly like the break-in forces themselves. If adequate defensive reserves of this type could be provided, however, it was found that equilibrium could be restored to the battle line. The 'blitzkrieg's' decisive break-through could be prevented, and the contest would revert to positional warfare.

After its first two years the Second World War continued to see some spectacular manoeuvring but it also saw a great deal of positional warfare and trench fighting. The tank by no means abolished this particular form of horror, as is often assumed, because it was found that the tank could be stopped. What happened, in effect, was that the development of tanks and anti-tank defences often cancelled each other out. The only really significant change was perhaps that the side which commanded the air could at last impose an important level of attrition upon the enemy lines of supply. At Alamein, Normandy and the Battle of the Bulge this was possibly the deciding factor in the outcome. Elsewhere it was important but less decisive. Heavy attrition still tended to take place in the front lines, just as it had during the First World War and in the Spanish Civil War.

Trench fighting in any war tends to be very like that in any other. Marshal V. I. Chuikov, for example, describes conditions in the Magnuszew bridgehead, August 1944, in terms very reminiscent of descriptions of the Western Front 1914–18:

As for what life in the trenches is really like, there is no need to say too much. Anyone who has no experience of it need only go down into a damp basement or cellar with a narrow slit of a window, and then imagine men staying in conditions like that for two or three weeks, or even several months on end, just waiting for their cellar-hole to be hit by a shell and themselves crushed under logs, boards, earth and mould. Added to which the soldier has to stand long hours of look-out duty, and has to sit in his listening post whether the rain pours or the sun blazes or the blizzard howls. And to crown all this comes that uninvited guest, the louse – who is the scourge of all under such conditions, and is no respecter of rank, title or honours. In Stalingrad they called the creatures 'tommy-gunners'; on the Northern Donets they were 'Vlasov's men'; and on the Vistula they were 'Faustniks'.[22]

In many of the Second World War battles the line became immobile for weeks and even months on end. In the Normandy fighting there was a fairly static front for almost two months; at Monte Cassino it lasted for six months; while at the siege of Leningrad it went on for two and a half years. In this war, as in previous conflicts, there were important 'static' phases as much as there were 'mobile' ones.[23]

In the First World War the fronts had been immobilised by machine-guns mixed in with the infantry positions. In the Second World War a sprinkling of anti-tank guns was added to the mixture to make it secure from armoured attacks as well as from infantry. The Germans were especially good at this, and led the way in making a flexible combination of all arms. Thus for them a 'tank battle' did not mean simply a duel between tanks meeting head-on, as it did for the British and other armies. Instead, it was to be a blend of tank manoeuvres, anti-tank gun fire and action by the other arms as well. It was to include both mobile and static elements as part of the whole. The Germans were particularly adept, for example, at using their tanks as bait to lure pursuing armour onto a screen of waiting anti-tank guns.

In the Western Desert the British took a long time to understand what was happening in actions of this type, since they had at first no such close relationship between their own tanks and guns. Far too often the British would push forward their armour in unsupported masses against anti-tank ambushes. Predictably, they would then find that the limited visibility from each vehicle and the difficulty of firing on the move made it almost impossible to locate and destroy the German guns. The attacker's losses tended to be high in these battles and sometimes had the effect of convincing British commanders that they were facing especially powerful enemy tanks, and not anti-tank guns at all.[24]

The Germans, by contrast, had quickly realised that a really heavy anti-tank weapon was essential. The redoubtable 8.8cm Flak gun turned out to be well suited to work of this kind, and was for some time much more powerful than anything the British were using in this role. Its long range meant that each piece could cover a wide area of front with its fire, thus allowing the principle of dispersion to be maintained without sacrificing mutual support. The open desert terrain lent itself to a defence by small, thinly-spread but powerful outposts built around this weapon.

We can here compare Rommel's ideas of 1941 with his earlier views from 1937. Essentially the general pattern is the same in each case, although the weaponry had certainly improved over the four years between the two. By 1941 he envisaged an increased dispersion of the outposts, an even greater stress on anti-tank defences, and a substitution of mechanised forces for infantry storm troop counter-attack elements:

. . . The outpost positions, up to company strength, must perforce be fairly far apart; but the whole line must be planned in adequate depth towards the rear.

Every defended point must be a complete defensive system in itself. Every weapon must be sited so that it is able to fire in every direction. I visualise the ideal arrangement of such defensive points on these lines:

One 88-mm. 'flak' gun should be sunk into the ground as deeply as the field of fire permits. From here trenches should radiate in three directions to three points – one a machine gun position, the second a heavy mortar position, and the third a light 22-mm. anti-aircraft gun, or a 50-mm. 'pak' [anti-tank] gun. Sufficient water, ammunition, and supplies for three weeks must always be available. And every man is to sleep prepared for action.

. . . In case of an enemy attack, the fire of our arms must completely cover the gap between the defended points. Should the enemy succeed in breaking through the gaps, owing say to bad visibility, every weapon must be in position to engage towards the rear. Let it be clear that there is no such thing as a 'Direction, Front', but only a 'Direction, Enemy'.

. . . The final decision of any struggle if the enemy attacks will probably rest with the Panzer and motorized units behind the line. Where this decision is reached is immaterial. A battle is won when the enemy is destroyed. Remember one thing – every individual position must hold, regardless of what the general position appears to be. Our Panzers and motorized formations will not leave you in the lurch, even if you should not see them for weeks.[25]

110

All this represents a pretty close translation of First World War methods into Second World War terminology. It shows a great continuity of doctrine which argues against any supposed 'watershed' in tactical affairs around 1940. As if to reinforce this point, we also find that wherever Rommel set up a fortified line – as at Alamein – it elicited a distinctly 'First World War' style of attack from his opponents. Thus at Alamein we hear of Montgomery's 'thousand-gun bombardment', his 'creeping barrages' and his infantry 'break-in' or 'crumbling' attacks. What else had Haig been trying to achieve at Passchendaele if not to 'crumble' the German positions in this way?

The Battle of Alamein also gives us a particularly vivid example of how a German tank attack could be stopped by infantry and anti-tank guns, once a suitably defensive mentality and guns of sufficient power had been deployed on the British side. This demonstration of the helplessness of armour came on 27 October 1942, just after the 2nd Battalion of the Rifle Brigade had occupied an advanced position near the point known as 'Snipe' on the western side of Kidney Ridge.[26] Just before dawn the (much understrength) battalion had dug a position of all-round defence into the soft sand, supported by nineteen of the new six-pounder anti-tank guns, including some from the 239th Anti-Tank Battery, R.A. The position, on a gradual incline, measured 900 by 400 yards, and was overlooked from the south-west by a low rise called 'Hill 37' and in close proximity, as it turned out, to the night leaguers of two groups of enemy armour. To the north-west was a part of the 15th Panzer Division, and to the south-west lay the 'Battlegroup Stiffelmayer', with a strength of about thirty-five tanks and self-propelled guns.

While it was still dark a detachment from the Stiffelmayer group had blundered into the battalion's position, and had been beaten off when two of their armoured vehicles were destroyed at very close range by the anti-tank guns. The Germans then took up covered positions about five or six hundred yards to the west and north. At dawn they broke cover to move away, but were engaged by the six-pounders and suffered no less than sixteen losses among vehicles which were presenting their vulnerable rear and side plates. There was a quick response by artillery fire upon the British position; but the scrub, the slight folds in the ground and the low silhouettes of the British guns combined to make accurate German observation impossible.

At this point the British received a rather hesitant reinforcement from 24th Armoured Brigade, which started badly by shelling the positions of its friends. As it came down from Kidney Ridge, moreover, the British armour attracted a lively fire from enemy tanks and anti-tank gun positions hull-down along the northern, western and southern horizons. The range was long for the British anti-tank guns to reply; but they succeeded in destroying three more enemy tanks. The intensity of the

111

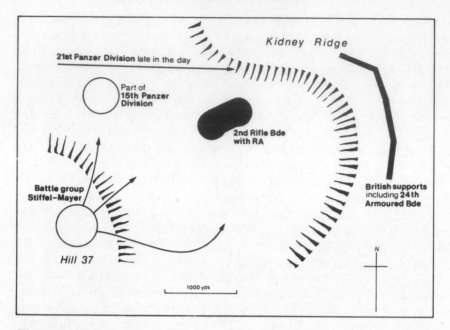

The action at 'Snipe'.

German fire, on the other hand, soon beat off the exposed British tanks, with the loss of seven of their number. They retired over the ridge to the east from whence they had come.

As the morning wore on the little garrison at 'Snipe' beat off an Italian infantry attack from the south, and then an attack by thirteen Italian M13 light tanks from the west. Immediately after this the British were able to engage the northern flank of a strong German armoured column moving east at a range of around 1,000 yards. Together with a cross-fire coming from 24th Armoured Brigade on Kidney Ridge, the anti-tank guns were able to destroy a further eight of these enemy tanks.

By this time many of the riflemen's bren carriers had been destroyed, and only thirteen of the six-pounders remained in action. In the early afternoon the Italian tanks from Hill 37 were able to make a fresh attack which came into the field of fire of only one of the British guns. For some time the issue hung in the balance; but in the end all nine enemy vehicles were destroyed by the single gun – the last one at a range of under two hundred yards. After this the enemy returned to shelling the British for several hours without making an attack, in which endeavour they were for a time joined – accidentally again – by some British tanks on Kidney Ridge.

At 5 p.m., however, the German 21st Panzer Division arrived to the north-west of the riflemen's position, and sent forward a group of forty tanks straight towards Kidney Ridge. They were apparently ignorant of the presence of the anti-tank guns a short distance from their exposed right flank, although they were soon to be disillusioned. The 239th battery found that it had four guns which could bear at ranges of between one and five hundred yards, and in a hectic duel they succeeded in destroying over twelve of the enemy Mark III's and Mark IV's. Perhaps more importantly, they persuaded the remainder of this imposing mass of armour to turn back and discontinue the attack.

The second wave of 21st Panzer Division next detached fifteen Mark III tanks to make a cautious frontal attack upon the British position. They made use of all the covered approaches they could, and came against a sector where they could be engaged by only three of the anti-tank guns. Fire was held by the defenders until the range had closed to a mere two hundred yards; but when it did come it quickly accounted for six of the enemy and beat the remainder back to a respectful distance.

After this, with an overall loss of well over fifty tanks, the Germans decided to call it a day. They had made casualties of about a third of the British garrison, had knocked out all but six of their guns, but had not defeated them. The gallant defenders were allowed to withdraw unmolested after nightfall, to spread the word that at last the Eighth Army had found an anti-tank gun which was actually capable of doing the job for which it was designed. As the historian of this action has said, 'the immediate lesson that was read to the whole of the army was that, when equipped with their own 6-pounders, the infantry could themselves see off a tank attack and inflict severe losses upon the enemy'.[27]

It was unfortunate for the Eighth Army that the Germans reverted to a generally defensive posture just at the moment when the British finally deployed an effective anti-tank gun. It was soon after this, moreover, that the first of a new generation of German tanks appeared in action: the Tigers and the Panthers. Compared with these excellent models the American and British armour was for a long time seriously deficient in both protection and gun power. In important respects, therefore, the attacking side now lost much of the edge which the Germans had seemed to enjoy in their early blitzkriegs.

Against consolidated defensive positions the attacker found that he was almost as powerless as had been his predecessors in the 1914–18 war. Progress through a zone of enemy battle outposts seemed to be as slow as ever, provided that the enemy's morale continued to hold out. Attacks were also liable, from time to time, to suffer the type of devastating check which the Germans had experienced at 'Snipe'. Given good interlocking fields of fire with powerful tanks and anti-tank guns, there seemed to be little for a defender to fear from an armoured attack. Given adequate petrol, also, a defender could still concentrate his reserves behind a threatened point and deliver counter-blows to regain lost ground.

That the allied armies did succeed in making slow but steady progress against the Germans says more for their material superiority, particularly in the air, than it does for any increased mobility which the tank may have brought to the battlefield. In the late summer battles of 1944, for example, British tank attacks in both Normandy and North Italy still seemed destined to meet bloody ruin against unsuppressed defences. Excessive faith in overwhelming numbers combined with insufficient co-ordination and control, and an unjustified demand for haste from the higher echelons, had all been depressing features of attacks in the First World War. In 1944 they appeared anew, now translated into the supposedly more 'modern' guise of tank warfare.

In the battle for the Gothic Line, for example, the 2nd Armoured Brigade was launched precipitously through the 46th Infantry Division on 4 September 1944, for what was intended to be the exploitation and pursuit of a beaten foe. This came after several changes of plan and two gruelling night marches which had deprived the tank crews of their rest. Some of the radios were inoperative because they had been netted onto the frequency of one of the BBC's light music programmes; and the brigade staff found they had no time even to make a reconnaissance. They were thrown headlong into a battle which turned out to be far from won, with inadequate infantry support and with a false idea of where the forward edge of the fighting had actually reached. To cap it all, some of the tanks encountered severe traffic jams as they tried to make their way forward to their start-line through the winding roads of the hilly Italian terrain.

One squadron of the Bays fell instantly into trouble as it moved up a valley:

As the tanks moved into sight, the German anti-tank gunners on Gemmano to their left and at Croce in front of them began to make good practice. The tanks came under steady fire from armour-piercing shot . . .

The road on which the tanks were moving had deep ditches on either side, and the big Shermans could only press on, returning the fire as best they could. The squadron rear-link radio tank, which was in contact with regimental headquarters, was hit at once, so that the tanks were immediately cut off from the regiment. One by one, three more tanks were lost, but the others succeeded in climbing the winding road to Coriano and the start-line. Here they were attacked by German infantry with bazookas, and another tank was knocked out and its commander killed. The tanks now came under heavy shell-fire, and as they crossed the start-line two tanks of squadron headquarters mistakenly turned away from the leading troop and took the road to Coriano by themselves.[28]

These two tanks fought a close action in the village, but both were soon knocked out.

All this seems to have been a far cry from the excellent theoretical precepts which the 2nd Armoured Brigade had been taught in their training. Only two months earlier they had been told that:

The infantry were the senior partners. They first decided how a position should be attacked, and were then given an appropriate number of tanks to support them. The rest of the tanks were responsible for helping and supporting the leading tanks from behind. The infantry scoured all bushes, hedges, and ditches to clear out any bazooka-men or snipers who might be in a position to pick off tanks or their commanders at a range of a hundred yards and less. The bazooka was a highly effective anti-tank weapon used by infantry . . . .

The Germans were very artful in its use and the bazooka-men and snipers sometimes held their fire until the leading troops were a mile or more beyond them. To overcome this it was necessary for the attacking troops to advance in great depth, so that when the leading infantry reached the objective their rear had only recently crossed the start-line. In this way the infantry and the tanks were well spread out over the ground just won and in a position to help each other to deal with the snipers. All-round observation in each tank was vital, because the enemy were just as likely to fire from either flank, or from behind, as they were from in front.[29]

The above passages serve to give us some idea of the difficulties which tanks could encounter in the face of infantry with anti-tank weapons. In the event it had turned out almost as pre-war analysts like Wintringham had predicted: the tank by 1944 was often no more than an auxiliary weapon of only occasional significance rather than the decisive war-winner its enthusiasts had hoped. Gone was the age of the successful massed charges which Guderian and others had been able to make for a short time at the start of the war. Once the defending forces had been fully alerted to this threat, their resistance had quickly stiffened. They soon made it obligatory for any tank advance to be properly escorted by infantry and other weapons – thus neatly reversing the earlier assumption that it was the infantry which needed the protection of the tanks.

The allied armies, however, took many years to recover from the trauma of 1940, insofar as they were slow to accept, in their heart of hearts, that tanks and infantry should always co-operate closely in battle. To them it seemed that excessive stress upon infantry support had been the biggest mistake which the ill-fated French armies had made. The British cavalry regiments, in particular, were long determined that they should not repeat the same mistake themselves. For many individual tank soldiers, on the other hand, it seemed to be folly to ignore the obvious aid which infantry and other arms could lend. Thus when Captain Geoffrey Bishop of the 23rd Hussars took up the advance in Normandy on 29 July 1944, he was happy to have an infantry escort with his tanks:

It is a fine morning, but there is a sharpness in the air, and the tenseness of a new adventure in a different sort of country, with these infantry soldiers to look after – although really they are here to look after us.[30]

115

By 1944 the shape of minor tactics had been complicated by a new generation of weapons, including many refinements to old ones. They tended to make it more rather than less dangerous to show oneself in front of the enemy, and the impression we have of the Second World War is of an increase in the importance of indirect fire at the expense of direct fire. Soldiers took more care to conceal themselves from the enemy than they had even in the trenches of the Western Front. The battlefield became 'emptier' than ever. Thus on one occasion in Normandy Captain Bishop found himself with a squadron of tanks holding a rather exposed field for a whole week. In that time he did not see the enemy more than half a dozen times, and did his best to conceal his own position in turn. He knew that the Germans were within a few hundred yards of his tanks, and there were occasional exchanges of shells or 'Moaning Minnie' (Nebelwerfer) salvoes to make the point. For much of the time, however, the area seemed to be completely deserted:

The same eerie stillness surrounds us, with only the occasional clatter of a bird's wing as it forsakes its roosting place in a hedge. Little movements of this kind become so noticeable under conditions of such tension – even the flutterings of the leaves left behind by the flapping wings acquires a significance of its own.[31]

The American combat historian S. L. A. Marshall made the same point:

The hardest thing about the (battle) field is that it is empty. There are little or no signs of action. Over all there is a great quiet which seems more ominous than the occasional tempest of fire.

It is the emptiness which chills a man's blood and makes the apple harden in his throat. It is the emptiness which grips him as with a paralysis. The small dangers which he had faced in his earlier life had always paid their dividend in excitement. Now there is great danger, but there is no excitement with it.[32]

Or again:

Even when the movement ceases and the opposing fire lines again become static, a reconnaissance along the friendly line is a point-to-point search for the hideouts of men, which is largely fruitless unless it is done by map. One not knowing where to search might move for miles along a main resistance line and see hardly a sign of war.[33]

The increasing emptiness of the Second World War battlefield was possibly a more significant change from First World War conditions than any development in armour. The chilling thing, perhaps, was that the landscape itself seemed to be much less seriously ravaged than it had been in 1914–18. Outside the towns battles took place in a countryside which, apart from the absence of living cattle, had almost a peaceful aspect. Shelling did not often continue for days on end, or with as many guns, as it had in the earlier conflict. To compensate, however, each

116

round had become more accurate, more lethal and more flexible. Better communications, in particular, now allowed front line observers to bring down accurate concentrations upon targets of opportunity without having to churn up the entire landscape at the same time. An individual walking in a field five miles away, for example, might now become a legitimate battery target. Such a sophisticated use of artillery had been generally beyond the reach of the soldiers of the Great War.

The mortar also partook of these developments, and was even more flexible than artillery because it was organic to infantry units themselves. The Germans were especially efficient in exploiting the quick reactions which this weapon made possible, and according to some estimates used it to inflict about a half of the total British casualties in the North-West Europe campaign of 1944–5. Mortar bombs were smaller than artillery shells; but their lesser velocity gave them a greater fragmentation effect because they did not bury themselves as deeply before exploding. Nor did they give as much warning of their arrival:

Mortaring is not like shelling, you don't hear the distant whistle of the shells coming slowly from afar. There is a sudden rush and a crack as the first bomb arrives, and you have hardly any time before the others land.[34]

The second most important cause of casualties was the mine. The Germans always left a good sprinkling of mines and booby traps behind them when they evacuated a position, and despite every effort to avoid them the pursuing troops would often suffer a toll of random casualties. This was a matter of pure attrition. It could not be in any sense decisive, but was undoubtedly extremely galling. On 14 November 1944 Major Martin Lindsay of the Gordon Highlanders wrote in his diary that:

There was a lot of talk this morning about the Hun having pulled out, as he was not seen by our patrols early today. I am glad this was not so as a few spandau boys are preferable to the legacy of mines and booby traps which he would certainly have left us had he gone back.[35]

Major Lindsay's diary forms one of the more vivid accounts to come down to us of the ten and a half months' fighting from Normandy into Germany. It conveys the great difficulties which even a half-defeated defender could throw in the path of an overwhelmingly superior army; and it conveys a sense of the attrition which the attackers had to suffer. In this campaign the Gordon Highlanders lost one and a half times as many men as they had started with on D-Day, and we gain the impression that most of these casualties were caused by shells, mortars or mines. Scarcely a day seemed to pass without an entry in Lindsay's diary referring to these problems. For example on 22 February 1945 he says:

Several more narrow shaves today. They wear one down in the end. Six months ago I found them slightly exhilarating, just as when one has ridden in a number of steeplechases without a mishap, it does one's nerve good to have a harmless fall. But now I have seen too much and have too great a respect for the law of averages. The shelling here has certainly been prodigious. Last night two new officers got hit on their way to join their companies . . . . We have been shelled, mortared and minnied all day, and the M.G. platoon has had twelve casualties out of fifteen men, from one unlucky shell, though of course they shouldn't all have been in one room.[36]

Contrary to the common impression that Second World War battles were easy, fast-moving and decisive affairs we find that they were in reality protracted, gruelling, nerve-racking and costly. There were more dangers to counter than there had been in the battles of a quarter of a century before; and it took a higher level of training and morale to overcome them. The tight-rope on which front-line soldiers walked had become thinner and less stable, reflected in higher levels of accidents, 'psychiatric casualties' and the general destruction of lives and property. War had become inexorably nastier. If there were occasional spectacular breaks by armoured spearheads, these always had to be paid for in full in the mud and blood of the close fighting which came before and after them.

Nor do we find that the widespread public faith in machinery was reflected in the writings of tactical analysts immediately after the war. Just as Ardant du Picq in the nineteenth century, and Tom Wintringham in the 1930s had identified machinery as the problem rather than the solution, so in the late 1940's we find a reaction against the easy optimism of those who believed that indirect approaches and mechanisation could remedy all ills.

In his book *Men Against Fire*, published in 1947, S. L. A. Marshall pointed to the confusion and moral isolation which surrounded the contemporary soldier in action. He saw clearly that this isolation was not reduced by piling up machinery on the battlefield, but greatly increased. Advances in weaponry and firepower did not serve to enhance the control which could be exercised over what actually took place on the ground, since 'The belief in push-button war is fundamentally a fallacy.'[38] Scientists and their sprawling industrial resources could certainly do much to heighten the intensity of combat; but they could not determine the ultimate victor. That was the responsibility, as it had always been, of the individual fighting man.

Marshall suggested that a new technical setting required a new type of soldier. It called for a man who could see through the bewilderments and anonymity imposed upon him by the machine, and who had both the will-power and the training to stick to his ever more complex task:

As I went about my work I came to see, more fully and more surely . . . that the great victories of the United States have pivoted on the acts of courage and intelligence of a very few individuals. The time always comes in battle when the decisions of statesmen and of generals can no longer affect the issue and when it is not within the power of our national wealth to change the balance decisively. Victory is never achieved prior to that point; it can be won only after the battle has been delivered into the hands of men who move in imminent danger of death. I think that we in the United States need to consider well that point, for we have made a habit of believing that national security lies at the end of a production line. Being from Detroit, I am accustomed to hearing it said publicly that Detroit industry won the war. This may be an excusable conceit, though I have yet to see a Sherman tank or Browning gun that added anything to the national defence until it came into the hands of men who willingly risked their own lives. Further than that, I have too often seen the tide of battle turn around the high action of a few unhelped men to believe that the final problem of the battlefield can ever be solved by the machine.[39]

It would seem that after every war, regardless of the much-heralded advances in technological aids, there will be a reaction of a similar type. After Waterloo there was a just appreciation of the power of the bayonet in the hands of good infantry. After Solferino and Mukden it was seen that although better firearms imposed looser formations and greater fire preparations, it was still the will to charge home which won the victory. After the bitter battles of 1918 the spirit of the picked storm trooper was recognised as an asset of incalculable value. And here, at the end of the only atomic war, we find the same lesson stressed once again. It was not the tank, not the gun; but the man. We are certainly entitled to predict that a similar lesson may be drawn from our more recent wars since 1945.

The *Corps de Chasse* and the *Mobile Group*:
Two Questionable Allied Responses to the
Panzer Division, 1941–1945

At the end of the previous two chapters we left infantry attacks baffled in the face of a wall of fire – first at New Orleans, where the British assault was bungled, and then at Sarrebourg-Morhange, where the French failed to apply the full depth of de Grandmaison's tactics. By 1945, however, we have found that enterprising infantry could once again find technique's to worm its way into an enemy defensive screen, and it was rather the tanks that were faced with a baffling wall of fire. They started off believing that they held the key to a completely novel style of warfare, but soon found that rather less had in fact changed than had been expected.

Just how novel were the conditions of combat between 1915 and 1918? For a long time they were not seen as such, since the old habits of thought took time to be jettisoned by tacticians, and of course trench warfare complete with grenades, tin helmets and bombproof shelters had already been venerable in the days of Vauban. Within the British cavalry, for example, the proud skirmishers of August 1914 were still optimistic for a breakthrough in 1915, although they became successively less credible – and individually less efficient – as the trench war continued. By 1917 Lloyd George's government was plotting to bring the cavalry home, and by 1918 even its genuine battlefield successes could be studiously mocked.[40] Nevertheless, it still had many champions, including Haig himself, and beneath the surface its offensive potential actually remained strong. New loose tactics exploiting fire and movement, both mounted and dismounted, had already been devised in the 1890s. When they were applied in the Great War, they allowed machine guns to be charged successfully throughout the hostilities – not only in Palestine or Russia, but even on the Western Front. One 1918 Canadian attack in France, for example, captured 230 prisoners, three cannon and no less than 40 machine guns. At Amiens in August two out of three charges succeeded and 1,300 prisoners were taken. In fact, cavalry units were very rarely massacred by fire in the way the popular stereotype would have it, but far more often deliberately kept out of battle by generals who either did not understand what they could do, or harboured deep-seated personal animosities against their commanders.[41]

Within the infantry platoon, equally, tactical attitudes changed only slowly. Hand- and rifle-grenades were added to the formal organisation as early as 1916, to all intents replacing the bayonet as the weapon of hand-to-hand combat. Nevertheless, assaults were still being made in excessively close-packed formations, despite several constructive experiments during the later phases of the Somme battle. It was not really until the very end of 1917 that new tactics were formally recognised at platoon level by the deployment of light machine guns, and a routine lightening of the assault formations. In the French army a similar reform took place as early as 1916,[42] but a profound change in outlook came only after the 1917 Nivelle offensive and the mutinies which followed.

In artillery tactics there was a similar story, since old-fashioned set-piece bombardments had to become bigger and clumsier before they could start to become more flexible and better adapted to the infantry's needs. On the Somme in 1916 the fire preparation lasted a week; at Passchendaele in 1917 it lasted a fortnight. It was only at first in more minor actions that bombardments were predicted without prior registration, and pared down to a few hours, in order to achieve surprise. It took time to perfect the full techniques of 'orchestration' in depth with the infantry, or to identify communications as a major problem needing a solution.[43] The old assumptions had to be 'tested to destruction' before they could be grudgingly replaced.

The slowness of change helped to emphasise the idea of a tactical 'revolution' in many people's minds, since the magnitude of the losses and frustrations had longer to sink in. It seemed obvious that the problem was a big one and needed a radical solution. Yet, ironically, the shocking indecisiveness of the Western Front was already fading, and tacticians were starting to meet the new challenges successfully, just as the wider public was starting to acquire the notion that battle was doomed to remain trench-locked forever. The traditional expectation that every mobile phase would be interrupted by sieges was too quickly forgotten, and too many strident interest groups[44] were allowed to drown out those who saw the Great War as merely a change of pace and style, rather than as a decisive rupture with the past.

The Great War threw up an overwhelming rhetoric of modernity and futurism, which led to some very grave consequences when it came to preparing for the next war. The most dramatic example of this perhaps lay in the role of strategic bombing, where technologically primitive air forces somehow managed to persuade the British and some other governments that they could destroy cities in a few hours, and that defence was quite impossible. The 1930s belief that 'the bomber will always get through' led to a massive diversion of funds away from battlefield weapons into strategic bombers which, by 1939, were vulnerable to fighters, still very inaccurate and almost negligible in their ability to damage the enemy war effort. It took the best part of five years' combat for bomber technology to catch up with the pre-war claims that had been made for it, during which time the men who had to 'go over the top' in this new battle of attrition were shot down like their predecessors in the infantry of the Great War. It is a bitter irony that the RAF could claim most of its many privileges by promising effective long-range attacks on Germany, yet in fact won its most obvious victory by destroying German bombers in the skies over Britain.[45]

If air forces were the most vociferous purveyors of over-optimistic futurism, the armoured forces were not so very far behind. In the mid-1930s the CIGS, General Sir Archibald Montgomery-Massingberd, followed most of army opinion in setting aside the horse and the purely colonial role in favour of a mechanised all-arms battle in Europe.[46] His reasonable and realistic views had a narrow tightrope to walk, however, since they were beset simultaneously from two sides. On one hand they had to beat off pessimists like Liddell Hart, whose fear of a new war and a new Passchendaele had led him into excessive reliance on standoff aerial bombardment.[47] This opinion was particularly influential with the Treasury between 1937 and 1939, when army funding was seriously cut. It meant that at the start of the war Britain had only two armoured divisions, and no factory capable of making modern heavy tanks.[48] Indeed, it would not be until March 1945 – long after the proverbial horse had bolted – that a really satisfactory main battle tank would eventually be fielded.

On the other hand, the reasonable middle ground was simultaneously under pressure from the Royal Tank Corps itself, where Fuller's disciples were preaching a futuristic 'all-tank' doctrine that left little room for the other arms. In Egypt Percy Hobart taught something like this to the nascent 7th Armoured Division until he was sacked for his obsession;[49] and similar views found their way into both the thinking and the official manuals of the whole armoured corps.[50] The idea was to get right away from the slow, complex and frustrating warfare of the trenches, and embrace independent manoeuvres in the open spaces – stylish, opportunistic and decisive. Fuller had sometimes described tanks as 'land battleships' operating like a fleet at sea,[51] while many were the subalterns who had thrilled to Lawrence's tales of desert raiding with the Bedouin. Classical theories of war were clearly obsolete in the new age of fast-moving, tracked steel cavalry!

The 'all-tank' ideology scarcely seemed to matter in the 1940 French campaign, where only one British armoured brigade could be effectively brought into action. On the contrary, the Germans' victory was often erroneously attributed exclusively to their tanks, which for good measure were also – scarcely correctly – assumed to be technically vastly superior to those of the allies.[52] Nor was emphasis on the tank pernicious in O'Connor's brilliant desert campaign, fought against the Italians over the following Christmas and New Year. He commanded acclimatised veterans who knew enough of their business to moderate dogma by common sense; so he was able to insist on the 'proper application of the three arms' in each of his battles. For example, the culminating fight at Beda Fomm was opened and sustained for the vital first four hours by 'Combeforce' – a small column containing armoured cars, infantry, anti-tank and field guns – but no tanks at all.[53]

After Beda Fomm, however, Rommel arrived with a solidly organised all-arms force, just as the British rotated their experienced formations out of the line. The brunt of the battle had to be borne by fresh troops starting almost from scratch, and basing their tactics only upon press speculation, theoretical manuals and rumour. Since the power of the tank and the romantic appeal of small raiding columns seemed to be central to desert lore, it was around those two features that most unit commanders wanted to centre their methods.

Despite considerable self-congratulation by both sides, however,[54] the supposed 'purity' of the armoured battle in the desert was ultimately an illusion. Rommel did admittedly enjoy an immense advantage over his compatriots in Europe, insofar as his command contained no horsed transport or gun teams. Nevertheless, his mobile spearhead of three armoured and two motor divisions was always counter-balanced by five or six infantry divisions.[55] This infantry was expected to be able to sustain itself without tank support,[56] in defensive positions or set-piece assault operations, leaving the armour free to fight its own opportunist war of manœuvre elsewhere. Within each Panzer Division, however, there was also the equivalent of one infantry and one artillery regiment, as well as the armoured regiment itself, making two to four infantry battalions supporting two tank battalions.[57] Hence the division contained adequate resources to undertake combats of every type and in every phase of operations – at least within a somewhat restricted frontage. German practice, furthermore, dictated that it should fight well concentrated whenever possible, with the artillery well to the fore.[58]

On the allied side the distinction between 'armour' and 'infantry' was seen in rather different terms. In essence, the tank was placed on a higher pedestal within both the armoured and the infantry divisions than was the case in the Axis armies: there seemed to be a quite widespread feeling that once friendly tanks had arrived on the battlefield everyone else could relax. The British maintained only around one active Armoured Division in the desert theatre until the start of 1942 – but it normally contained an enormous number of tanks. In the 'Crusader' operation of November 1941, 7th Armoured Division put no less than 480 tanks into the field, split into three armoured brigades each of three

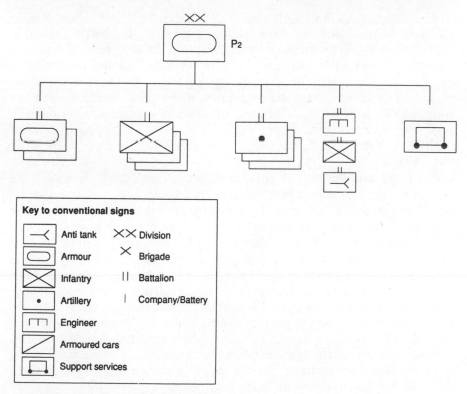

21st Panzer Division, November 1941.

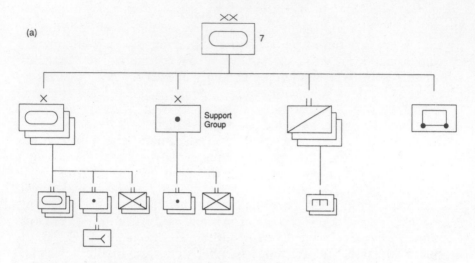

7th Armoured Division, November 1941.

tank battalions. Each brigade thus had more tanks than a whole Panzer Division. Supporting all these tanks, however, the British division contained only three infantry battalions and little more artillery than could be found in a German division. From these we may further discount the divisional 'support group' since, in the dispersed penny-packet fights so dear to the desert cavalry, it often fell behind the armour and missed the battles. The level of infantry and artillery normally available to the British tanks can therefore be described as almost negligible. Each brigade contained only about two infantry companies and 16 field guns.[59]

As if to compensate for this 'tank-heavy' armoured division, the British would normally attach independent, heavy but slow-moving, 'I' tank[60] brigades and regiments to their infantry formations. Unlike their opponents, they were not usually happy for infantry to fight without any armoured support at all, but instead tended to give it more than the case actually warranted.[61] This habit may be attributed not only to the inflated prestige enjoyed by the tank itself, but also to the relatively poor quality and quantity of the towed anti-tank artillery available to the British before the summer of 1942. A notorious despondency followed any removal of close-support armour, since the Afrika Korps had overrun all too many British infantry formations for there to be any complacency on that particular point.

Diametrically opposite to the Germans in their concepts, therefore, the British fielded armoured formations, for a 'cavalry' role, that were highly mobile but terribly fragile – while their infantry divisions, if supported by a good 'I' tank brigade, could be expected to achieve almost as much as the Panzer Divisions, albeit at a much slower pace. The Germans thus placed most of their emphasis on their armoured spearheads as opposed to their infantry divisions, whereas it was the infantry division which tended to be seen as the predominant formation within Eighth Army.[62] For the British it was to be a characteristically 'infantry' style of warfare which emerged from the desert campaign, and it was to be Montgomery the methodical infanteer who became their greatest captain.

This is not to say that the British particularly wanted the war to become dominated by infantry, since we have already glimpsed their initially deep faith in the tank and dislike of static attritional combat. Their problem was rather that in the eighteen months following Beda Fomm in February 1941, the tank did not seem capable of delivering a convincing offensive. Once Rommel had established siege lines around Tobruk, he proved frustratingly difficult to dislodge. When he was finally pushed back in December, it was only after a distinctly Pyrrhic[63] British victory in the 'Crusader' operation, which was soon reversed in humiliating disaster at Gazala. This in turn led to defensive battles at First Alamein and Alam Halfa, and it was only in October 1942, at Second Alamein, that the Eighth Army began to roll convincingly westwards. Since it had enjoyed at least a three to two superiority in tank numbers throughout most of the period, it could surely be forgiven for believing that this particular weapon was something less than the total key to successful offensive action that had originally been promised.

Montgomery's contribution was to condemn panaceas and remind the army of the continuity of tactical principles, the 'proper application of the three arms', and firm centralised control. Out of the window went

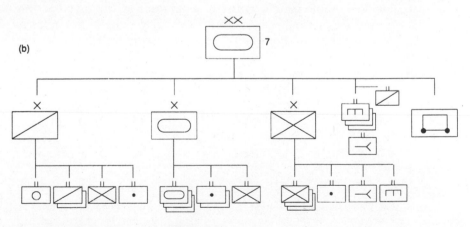

7th Armoured Division, November 1942.

the last vestiges of the all-tank armoured division.[64] Out went the freedom of formation commanders to pick and choose their manoeuvres as they wished. Out went the tiny independent 'Jock columns' that used to dash around the desert looking impressive but – after Combeforce's epic stand at Beda Fomm – rarely achieving a great deal. Montgomery even reversed Auchinleck's quite sensible desire to fight by brigade groups, which had performed well at First Alamein and elsewhere. He believed that they contained inadequate supporting arms and services and, by neglecting the principle of concentration, led to defeat in detail. He therefore designated the Division, united in one place and with all its assets available, as the basic tactical unit.[65] This was a more cumbersome solution than Auchinleck's, and did not long survive in practice; but it was intended to ensure that centralised control could be maintained and sufficient resources brought to bear in each engagement. Distressingly, however, it also helped to move combat back to the type of protracted frontal slugging that had been so roundly reviled in the Great War. In its new guise, this could be made almost acceptable only by a realistic promise that victory would eventually be inevitable.

Montgomery always wanted to reassure his men that everything was totally under control. His approach drew on the old infantry principle of imitating a parrot climbing the wall of its cage – with two feet always secure before the next forward reach with the beak. He certainly wanted to seize the initiative and force the enemy into predictable reactions;[66] but within that framework he would deliberately set realistic targets that his men could attain at relatively low risk. This approach worked well in ordinary circumstances, but was less successful when it came to moving predominantly armoured formations forward into speculative ventures on the far side of predominantly infantry battles. No less than his predecessors, he seems to have retained a rather rigid view of how 'armoured' action should be conceptually separate from 'infantry' action. As in the Great War, in fact, he wanted the cavalry to be kept back as a *corps de chasse* or *masse de manoeuvre*,[67] ready to spring blithely through 'the G in Gap'[68] once the gap had been made. Also as in the Great War, he encountered serious practical problems of timing, logistics and traffic control when that sublime moment actually arrived. After experiencing this, Haig had called for his cavalry to be more closely integrated as a rearward portion of the assault itself, but this advice had not been heeded either at the time or later in the Second World War.[69] The legacy was that at Alamein the armoured formation commanders who were passed forward through the infantry

. . . feared that their tanks, on emerging from a minefield, would come under effective fire from anti-tank guns sited for the purpose and dug in. In fact this happened again and again, for the British armour frequently failed to make full use of the supporting arms.[70]

126

At Alamein the armoured corps was at first even given a defensive role upon emerging on the far side of the infantry battle. They were supposed to cover the 'crumbling' of the German defences, in rather the same way that the Egyptian anti-tank screens were deployed in 1973 to cover the reduction of the Bar Lev line. They failed to give full satisfaction, however, and later in the battle they were pulled back,[71] regrouped and switched to passing through a new assault – 'Supercharge' – which also at first fell short of a total breakthrough. The *corps de chasse* was finally able to move into full pursuit only after the Germans had started to retreat – but then it rapidly fragmented, ran out of supplies and was stymied by the enemy rearguard. Even though Montgomery would be able to claim an impressive mobile victory when X Corps outflanked the Mareth Line in May 1943, his general experience of massed armour at Alamein had been less than happy. In January 1944 he even flatly stated 'I will never employ an armoured corps',[72] which seemed to preclude any further use of a *corps de chasse*.

This resolution did not long survive D Day, however, and in Normandy there were several new attempts at a massed armoured breakout. Nevertheless, the pre-planned infantry and artillery preparations usually failed to go deep enough, leaving the enemy with plenty of time to reinforce defensive screens further to the rear. Operation 'Goodwood' was perhaps the most spectacular example of this, but there were others.[73] Tanks were

too often . . . exhorted to dash forward when everything else had gone wrong and the main enemy defence had hardly been touched. The result was that when opportunities really did occur and when the opposition really was weak, the infantry commanders were not prepared to put their trust in exploiting tactics, preferring to plod on with successive "well laid-on" attacks and many of the tank soldiers by that time were chary of doing it anyway.[74]

The final breakout came only in operation 'Cobra', on the American front just south of St Lô, where by 27 July two days of carpet bombing and intense infantry attacks had almost worn a gap in a much weakened German defence which lacked its customary depth. Flying in the face of many earlier experiences, General Collins now successfully gambled that the gap could be fully opened by an armoured assault. His tanks started to roll, aiming to halt and consolidate at Coutances; however, in the event it was found that Patton's VIII Corps could pass through, exploit in great depth and take on the full mantle of *corps de chasse* that had so long eluded the British.[75] Patton was able to motor through France, pre-empting battle as the Germans had done in 1939–41, until finally bogging down on the fortified fields of Gravelotte, outside Metz.

It has often been stressed that Montgomery was 'cautious and methodical', but in fact that is not the whole story, since in most of his battles he initially set a very optimistic and demanding timetable for the attack. Although he correctly estimated that the whole battle of Alamein would last 10–12 days, he expected the main enemy position to be breached in the first night. The armour intended to move through the gap, to cover it from the far side, was set in motion only four hours after the initial offensive began.[76] Again, in both operation 'Market-Garden' and the battle of the Reichswald, XXX Corps' spearhead had to pass down a single long road within three days fom H Hour, when in fact it needed, respectively, a fortnight and a month.[77]

In Normandy Montgomery was more ambitious still, since he wanted Caen to be seized on D Day itself, as part of an almost instantaneous transition from infantry landings and beach-clearing to a phase where 'great armoured columns' would sweep inland 'to seize objectives and to keep the enemy unbalanced'.[78] This was partly because everyone, especially Churchill, had learned from Anzio that it is vital to move off the beaches at top speed;[79] but it was also a far from unique instance of Montgomery trying to overrule difficulties by willpower alone. Anzio had highlighted some very important potential sources of delay in amphibious operations, but they were no more allowed to modify the CiC's timetable than would other predictable obstacles at 'Goodwood' or 'Market-Garden'.

Montgomery's reputation for caution is surely based not on his original intentions in these battles, but on the painstaking slowness with which they actually unfolded. It was one thing to designate ambitious objectives, but quite another to achieve them quickly. Yet we do not think of Haig as 'cautious' when, although he ordered and expected a breakthrough on the Somme on 1 July 1916, his battle actually continued until half-way through November. We tend to think of his long battle more as an act of recklessness or waste. We are suspicious of protestations that it was constructive after all, just because it 'wrote down' the enemy strength or helped our allies to make progress elsewhere. Where, then, is the secret of Montgomery's reputation as an effective commander in the offensive? Presumably it lies in the fact that each of his battles – apart from 'Market-Garden', for which he *is* accused, precisely, of recklessness – eventually led to a significant bound forward. This allowed the defeats, delays and frustrations of the fighting to be accepted after the event as intended parts of the plan. He himself systematically sought to reinforce this impression, and to rewrite the history books accordingly.

When he first took over Eighth Army Montgomery had been very anxious to show that, unlike his predecessors, he would not be caught out by one of Rommel's surprises. He was also keen to give his soldiers the self-confidence of believing that everything was going strictly according to plan. Preparations for the battle of Alam Halfa were based on this,[80] from which it was but a short step to a doctrine that every battle should be a predictable, carefully organised and timetabled event. Hence it was, ironically, Montgomery himself who drew his own caricature as a slow-moving and cautious general. It came to be vitally important to him to claim that he always kept 'balance' in his battles, which he disarmingly defined as a state whereby 'it is never necessary to react to the enemy's thrusts and moves'.[81]

Such a theory of tactics would seem quite astounding to most students of war, but it was a conceit that Montgomery found possible because his numerical and material superiority made the ultimate outcome of his battles almost a foregone conclusion. His timetables, however, could not be so guaranteed, least of all for the mobile armoured spearheads which he so often threw forward recklessly in front of his more methodical infantry. In these cases he could easily have explained away the resulting loss of time and casualties in terms of 'risk-taking' or 'the accidents of war', but instead he would always insist that they had been 'part of the plan from the start'.[82] Since he felt psychologically unable to admit to gambling, improvisation or swift changes of plan – all qualities which are surely close to the heart of armoured warfare – a genuinely mobile and fluid battle became officially unthinkable in his army. 'Adjustments to maintain balance' could be acknowledged, but nothing less predictable than that. The very concept of armoured warfare thus fell out of favour in Britain, in a way that would have seemed utterly backward-looking to the daring modernists of 1939–41. Astonishingly, therefore, the military outlook of a whole generation of British officers came to be turned through 180° by little more than the style in which one individual chose to write his own war memoirs.

Montgomery could not, however, suppress the reputation of the Germans in mobile war, which survived even their comprehensive defeat in every theatre. For a variety of good and less good reasons, western commentators from 1940 onwards have taken a perverse delight in praising their foe. Admiration for the German soldier – particularly as he appeared conveniently close at hand in Normandy[83] – has become something of a cult among students of tactics. It may be worth remembering, however, that the *bocage* south-west of Caen was 'a defender's paradise'[84] which did not, perhaps, need very exceptional qualities to hold. Certainly, the Germans do not seem to have done as well in their counter-attacks as the allies did in at least some of their attacks, while the very uneven standard of their manpower gives ample credence to the following opinion from one British infantry platoon commander. He found few high military qualities among the German soldiers he met, despite acknowledging their generally superior weapons and tactics:

In many attacks the prisoners we took outnumbered our attacking force and German units who would continue to resist at close quarters were few indeed. Unlike us, they rarely fought at night, when they were excessively nervous and unsure of themselves. Where we patrolled extensively, they avoided it.[85]

For the war on the eastern front, however, it is harder to find dissenting voices from the view that the Germans were greatly superior to their adversaries.[86] This is partly because the accessible uncensored sources on the campaigns in Russia have come largely from the German side. In the absence of Hitler's own point of view, furthermore, it is the specifically military interpretation of events that is predominant. A number of his generals have brought back confident, action-packed memoirs of their adventures in the open steppes, telling a tale guaranteed to stir the heart of young NATO officers. In Russia the German commanders of armour could successfully realise the free, virtuoso manoeuvres which had eluded their British contemporaries in the desert. Small detachments of tanks could switch off their rear link radios[87] and follow their own initiatives, wheeling round behind cumbersome enemy groupings to launch devastating assaults upon flanks and rear. Thus Balck's depleted 11th Panzer Division on the river Chir, during the attempt to relieve Stalingrad, was often cited as a model of a 'fire brigade' rushing to extinguish each enemy break-in before it had been consolidated. Manstein's operational counterstroke at Kharkov represented a similar tactic on a larger scale, and a series of other variants are regularly paraded.[88] These actions have come to be seen as the classic reference points of manoeuvre warfare in the era after Fuller.

From the Russian point of view, however, the picture looks very different. The relatively low fighting quality of their individual soldiers can to a large extent be explained by the incorrect political line of J.V. Stalin in the years before the Fascist assault of 22 June 1941. He deprived the army of its best trained leaders by purging the officer corps, and allowed most of the tanks and aircraft to be taken by surprise in exposed forward positions where they were rapidly destroyed. This was soon followed by the loss of many war industries and the displacement of others across the Urals. From there an army had to be improvised from almost nothing, and it was near miraculous that such a big and successful one could be mobilised as fast as it was. Whereas the Germans could usually field around 200 divisions in Russia, by mid-1943 there were almost 500 Soviet divisions to oppose them.[89] The important point was therefore not that these formations suffered from defects in detail or tactical finesse, but that they could fight effectively at all.

The Russians survived because they remained undaunted. Not only did they launch major counter-attacks against the initial onslaught, but they did so on an even greater scale over the winter of 1941–2, and then at Kharkov in the spring. In the autumn they were counter-attacking again at Stalingrad, and then once more at Kharkov, Donbass, Orel and Proskurov-Chernovits during the first half of 1943. It has to be said that in none of these big counter-offensives were the Soviets particularly successful – except at Stalingrad itself – and the lessons learned were mainly of the 'how not to do it' variety. In common with the Anglo-American experience in the Mediterranean, in fact, it was only during 1943 that a truly confident and successful combat style could eventually emerge. Nevertheless, in the Kursk and Belgorod-Kharkov operations during the summer of 1943, the Russians began to show just what they had to offer.

An important difference between the Russian and the Anglo-American experience was that whereas the latter often seemed to make up their tactics as they went along, the Russians were able to draw upon a detailed official doctrine for modern warfare that had been mapped out by Triandafillov and Tukhachevskiy, and incorporated in the 1936 field regulations. The 'deep battle' with an all-mechanised, all-arms force was envisaged, including 'manoeuvre tanks', 'mechanised cavalry' and mechanised airborne forces simultaneously to disrupt the enemy's rear and exploit breakthroughs in his forward positions.[90] More specifically, a third of the Soviet forces would pin the enemy on a broad front as the remainder acted as *shock armies* breaking through in depth on a narrow front, opening a path for *mobile groups* of horsed cavalry and mechanised troops, with airborne support, to exploit far to the rear. The mobile groups would encircle the enemy to a depth of as much as 2–300km, leaving him to be mopped up by a second echelon of the less mobile forces.[91]

By the end of 1943 the Red Army had been re-organised several times since the 1936 operational doctrine had been laid down, but its final shape was nevertheless very well fitted to carry that doctrine into effect. The bulk of the army, as with the Germans, was made up of infantry divisions and armies which relied mainly on horse-drawn transport. Unlike German infantry divisions, however, these formations were notoriously cumbersome, inflexible in their tactics, and poorly equipped with supporting services apart from artillery. They did often fight at night and could make successful infiltrations, and in the last two years of war they were regularly topped up to strength in a way that the Germans could not be. Nevertheless, they were essentially cannon fodder, for whom the highest military virtue was to act with cog-like uniformity and predictability in the great machine of war. They provided the model from which most western caricatures of the Red Army have been drawn ever since.

In clear contrast to the caricature, however, stood the manoeuvre elements composed of tank, mechanised and horsed cavalry corps. The tank and mechanised corps were the mainstay since, with around 200 tanks at full establishment, each of them was roughly equivalent to a strong Panzer Division. A tank corps further contained over 4,000 motorised infantry (including an SMG battalion attached closely to each 'brigade' of 60 tanks) and 152 mortar or artillery tubes, while a mechanised corps had more than 10,000 infantry and a massive 252 tubes. To give extra flexibility in difficult ground for vehicles, a cavalry corps might be added to a mechanised one, and in practice these apparently anachronistic mixed formations often did very well.

The mobile forces were given the best available equipment and, still more importantly, officers. Unlike their colleagues in the ordinary infantry divisions, the tank and mechanised infantry commanders had to think on their feet and, within certain limits, adapt their battles to changing circumstances. This applied particularly to the higher echelons of command, where actions at the operational level were planned. The Russians were quite well aware that man for man their troops were only something like a third as efficient in tactics as the Germans, and that there were less than three times as many of them. Victory, therefore, depended on cleverness in the levels higher than tactics. If the majority of local combats were predestined to be lost, it was essential to ensure that a few really important ones would be won, and then transformed into a grand encirclement of the enemy that would quickly make his tactical superiority irrelevant.

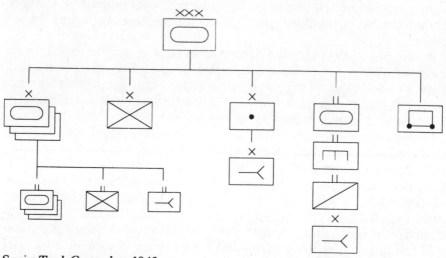

Soviet Tank Corps, late 1943.

The hallmark of the Russian view of warfare was an emphasis on co-ordinated large-scale action in depth, using 'operational art'. This was a very deliberate system during the vital phases of preparation, strategic deception-cum-concealment (*maskirovka*), and initial break-in. Provided the troops could be lined up and kept hidden, their chances of success could be precisely calculated. For example, specific ratios of artillery per kilometre of breakthrough sector could be relied upon to open a gap. This figure was between 103 and 220 tubes for the Stalingrad encirclement, rising to between 277 and 420 tubes for the Vistula-Oder operation.[92] Again, there were specific rules for selecting the moment to commit the mobile group. This had been found ineffective when delayed six days in the disastrous 1942 Kharkov offensive, but totally successful when released during the first day of the Stalingrad encirclement. The normal expectation came to be that it would be released as soon as possible, but certainly within the first 36 hours.[93]

The massed mobile breakout itself, although open to flexible changes as the battle progressed, could also be carefully planned in its general shape, to the extent that by the end of 1943 it was felt that, with sufficient planning and *maskirovka*, a deep penetration and even encirclement could almost be guaranteed. The difficult part of the battle was now thought to consist in securing the claims staked out, against attack from both within and outside the lines of encirclement. The mobile groups therefore became heavier and more sustainable at longer distances from their bases, with each 'Front' putting out a united formation made up of between one and three homogenous tank armies, each of several mobile corps. This force would be expected to complete the breakthrough started by the infantry armies, although it might well lose a third or more of its strength in the process. Then it would advance in a concentrated mass, avoiding over-dispersion, to effect an operational encirclement and repulse counter-attacks. In addition to the Front's mobile group there would also be local mobile groups for each infantry army, consisting of one or two tank or mechanised corps operating in a similar but smaller way.

All this was obviously conceived on a vastly larger scale than Montgomery's attempted breakouts between Alamein and the Reichswald. Where he almost always had less than a dozen divisions committed to a breakout, the Russians would regularly have many more than that in each Front, and often several Fronts in each operation. Where he might think it daring to yoke two armoured divisions together in a single *corps de chasse*, the Russians could call on up to a dozen homogenous tank armies, and thought it normal to use two or three of them grouped together on a single axis. Where he believed 1,000 guns made an exceptionally heavy barrage, the Russians were habitually using 2,000 or more for their breakthroughs, and at Berlin they had over 6,000. Where Montgomery often took weeks to achieve his initial breakthroughs, they took just hours, and would realistically hope to close down each whole operation after only ten days – but some 2–300km further to the west.[94] Perhaps most telling of all, the Russians had far more experience of the whole process of operational art, and far more opportunities to learn by their mistakes. Where Montgomery may have mounted a dozen or so offensives at most, they claimed to have mounted no less than 250 Front operations. At first these suffered from all the setbacks and frustrations that were familiar to the western allies, but in the last eighteen months of the war they were being properly set up and running smoothly in a way that was quite unknown in the west.

Despite all these important differences of scale, speed and professionalism, however, Montgomery and the Russians found themselves working to a not dissimilar basic conceptual pattern. They were both very systematic, and planned their build-ups, *maskirovka* and artillery preparations with great care and deliberation. They both aimed to launch a mobile group to complete and then exploit a gap in the infantry battle at an early stage of the fighting, and were prepared to accept heavy casualties in order to do it. Having learnt by hard experience to distrust the tactical sureness of their troops, they thought it important to over-insure against accidents. They both believed that the best hope of success lay in being so overwhelmingly strong at the points where it mattered that there would be nothing the enemy could possibly do to turn the blow. Both of them wanted 'balance', or the ability to transcend the accidents of minor tactics. In their own doctrinaire and statistical manner, in fact, the Russians were no less insistent on the inevitability of their timetable than Montgomery himself. Incongruously and unexpectedly, but nevertheless distinctly, we may even detect a passing echo of his historiographical style in their official accounts of these remorselessly victorious campaigns which won the war.

The question we should perhaps ask, therefore, is whether the operational art, as glimpsed by Montgomery and perfected in the Soviet Union, was really the only possible approach to offensive warfare in the conditions of 1943–5. Could not smaller but cleverer and more independently manoeuvrable spearheads have achieved as much but more economically? Could not the opportunism shown by such panzer leaders as Rommel or Balck have been harnessed to a lightning war which maintained the impetus won in 1939–41, even against the newly mobilised opposition? Could not a new style of warfare have emerged that would have been less elephantine than the *corps de chasse* or the Mobile Group, and nearer to the vision of the pre-war futurists?

We cannot know the answer to these questions, which must remain among the imponderables of history. The Germans ran out of resources at the moment when they might have demonstrated a fully mature version of armoured warfare, but not before they had severely mauled their opponents. Both the western and the eastern allies found they were left with no alternative but to improvise a system that worked, using only the relatively unpromising elements that had survived the first onslaught. They both therefore chose a cautious system that relied on over-insurance rather than fluid or brilliant manoeuvres. The liberation offensives may have succeeded in reaching Bremen and Berlin, but they never entirely convincingly shook free from the mud and the blood into the green fields beyond.

5

1965–73: The Alleged Supremacy of Technology in Vietnam

The Second Indochina War was the biggest and most important war between 1945 and the present day. At the political and psychic level this truth has been well grasped; but in purely tactical affairs it has not. There is a widespread belief that Vietnam was somehow outside the main line of development in the Art of War. It tends to be dismissed as either a sideshow or an aberration.

There are many reasons for this neglect. The military lessons of Vietnam seem to relate to something very different from the classical European style of combat. It was a war fought in an exotically un-European climate and terrain; with an 'internal security' or counter-guerilla task; and with an emphasis upon infantry rather than armour. There was a heavily political element which often seemed to undermine the military operations; and to cap it all there is a persisting bad conscience about Vietnam which frequently interferes with objectivity in debate. Small wonder, then, that the military lessons of this war have been eclipsed today in rather the same way as were those of the American Civil War during the late nineteenth century.[1]

A further problem is that the war in Vietnam appears to be dauntingly complex. It defies easy study, since it was actually made of up five closely-interwoven and concurrent wars: *viz*. the international political struggle, the strategic bombing campaign, the interdiction campaign, the mainforce battle, and the pacification or 'Village War'. None of these five campaigns had a clearcut outline, since each to some extent overlapped all the others. A great deal of confusion has resulted from this close proximity of the five different campaigns.

In order to look at the tactical developments in Vietnam, however, we must first attempt to isolate the mainforce war from the other four. For the purposes of the present study, at least, we should try to forget about

136

the politics, the strategic bombing campaign and the village war; and should concentrate solely upon the pitched battles between regular mainforce battalions and divisions. This will enable us to examine combat techniques as such, and incidentally help us avoid the morass of political controversy which so bedevils the study of the war as a whole.

The mainforce war has certainly attracted plenty of argument; but this

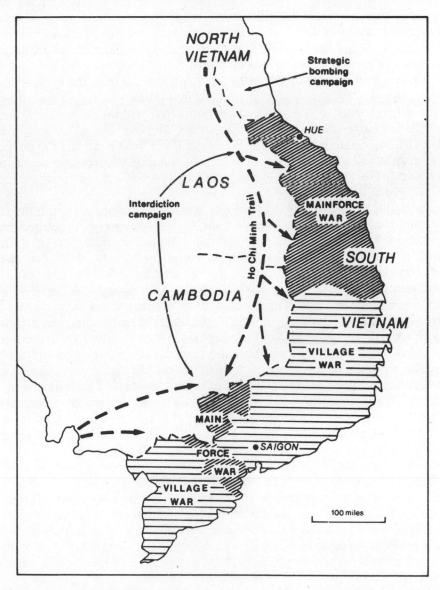

The four wars in Indo-China, 1965–73.

137

has tended to be about morality or strategy rather than about tactical efficiency. In a curious way both sides in the Vietnam debate seem to agree that in battle the American forces were very effective indeed. Both supporters and opponents of the war stand in awe of the mobility, firepower and futuristic electronic instruments which were used. Perhaps it is true that the enemy was hard to find, or to distinguish from friendly civilians; but once found there seems a wide consensus of opinion that he could be dispatched very quickly. In this sense the mainforce war has been surprisingly uncontroversial, although it is possible that the consensus itself is misplaced. We will return to examine its basis in greater detail at a later point.

The mainforce war has also been less controversial than the village war because it was fought largely in uninhabited border areas.[2] It spilt over into only relatively few of the heavily populated provinces; mostly in the northern quarter of the country and in the Mekong Delta. Only in the Tet Offensive and in some of the battles of 1972 did it enter the towns. In the main it had the character of conventional jungle warfare between two regular armies, with few of the complexities occasioned by the presence of civilians.

In both the mainforce and village wars the communists employed guerilla tactics; fighting a war without fronts, moving in small groups and using all the vegetal and human camouflage they could. This similarity of tactics has sometimes led commentators to imply that the two wars were essentially the same; but in fact they were not. The village war was a local affair, run mainly by the Viet Cong and depending heavily upon Mao Tse Tung's principles of protracted revolutionary war. The mainforce war, by contrast, was an externally organised attack by the North Vietnamese Army. It depended far less upon the dynamics of political subversion than upon the traditional military principles of regular armies. For the village fighter there was essentially no alternative to fighting as a guerilla. It was inherent in his relationship with his village and his local political cell. For the mainforce fighter, by contrast, 'guerilla warfare was a tactic, not a genre'.[3] It was a military stratagem adopted by choice, which could be abandoned, if circumstances were appropriate, in favour of more overt or close-order fighting.

If we take the mainforce war in isolation, therefore, we will be reflecting an important reality. We will also be following in the footsteps of General Westmoreland himself, since he seems to have done his best to keep American troops away from the pacification effort in the villages. His ideal was to concentrate the Americans for large-unit battles in the uninhabited border areas, while the South Vietnamese forces looked after the 'internal security' of their own people. He thus decreed that there should be 'two wars' within South Vietnam, and not just one.

The rationale behind Westmoreland's strategy was twofold. In the first place it inserted a screen or 'shield' of American troops between the Viet Cong in the coastal villages and their NVA colleagues coming in through Laos and Cambodia. This screen would 'hold the ring' for the South Vietnamese pacification effort in the villages. Secondly, Westmoreland aimed to inflict prohibitive attrition upon the NVA so that it would be dissuaded from further adventures in the south. For political reasons the Americans were by and large prevented from making decisive or direct attacks upon NVA bases beyond the frontiers of South Vietnam itself; so the only option remaining was the much less decisive one of attrition. By killing as many NVA soldiers as possible within South Vietnam, it was hoped that the northern leaders could eventually be persuaded of the futility of their campaign, up to the point at which they would call it off.

In the event Westmoreland's policy seemed almost to have been mirrored by the NVA's own pattern of operations. As the Americans built up their strength between 1965 and 1968, so did the NVA. Northern formations grew in both size and sophistication, and sought to fight big battles of an ever more conventional nature. They still tried to pass reinforcements to the village guerillas; but this increasingly became secondary to the task of fighting big battles in the border areas. It was a prime aim of the NVA to inflict as many casualties as possible on the Americans, and they were quite ready to meet Westmoreland's mainforce challenge with a mainforce of their own.

It is arguable that Westmoreland's emphasis upon the mainforce war was a cardinal error, insofar as it diverted American energy and resources away from the ultimately critical task of building up the South Vietnamese. The village war tended to be neglected, and it has been calculated[4] that less than 10% of total American resources were allocated to programmes relating to it. It was only after the Viet Cong had suffered heavy casualties in the 1968 Tet Offensive that progress could be made in this field. Tet shocked the South Vietnamese into setting their house in order, so for the first time some of the vital but much-misunderstood lessons of the Malayan Emergency came to be applied to pacification. Another result of Tet, however, was to precipitate American withdrawal. This meant that just as they seemed to be making genuine progress, the Americans had to change gear and belatedly hand over the mainforce war to the South Vietnamese. Despite early signs that this could be successfully achieved, it was eventually shown to be an unrealistic policy. The final victory of the North in 1975 was the result of a South Vietnamese collapse in the mainforce battle, following denial of US support.

The reason why the Americans failed to build up South Vietnamese forces in good time was that they were distracted by their own mainforce war. This war was far from a walkover, and we may be entitled to question whether the supposed battlefield supremacy of the Americans

139

was really as great as is usually assumed. It did, after all, take them over three years to win their mainforce victory. It was only around 1969–70 that they had genuinely defeated the NVA inside South Vietnam. Surely this is considerably longer than they had hoped, much longer than they predicted at the time, and longer than is normal in conventional wars elsewhere. What went wrong?

There are many structural factors which the Americans can invoke to explain their relatively long delay in mastering the enemy mainforce. It took much longer than they had predicted to build up their logistic and support facilities, and it took longer than they had hoped to obtain the release of sufficient combat units. There were a number of other political decisions which may also have impaired efficiency – e.g. the one year tour of duty, the refusal to send reservists to Vietnam, and the insistence on a high standard of comfort in the rear base areas. The North Vietnamese, for their part, continued to send large formations south long after they had been expected to desist. Although they never remotely approached the total numbers of free-world troops in South Vietnam, they could often field as many men in mainforce manouevre battalions. It has been calculated that the true ratio of 'foxhole strength', excluding logistic or auxiliary troops, fluctuated between 1.2 and 0.7 free-world soldiers to every one NVA.[5] The Americans and their allies did not therefore enjoy a significant numerical superiority in the mainforce battles.

It is usual to claim that despite all these deficiencies and delays, American forces nevertheless exhibited a devastating ability to smash enemy formations wherever they were found. Both sides in the debate seem to accept that, as far as the purely mainforce battles were concerned, the Americans had more than adequate skill and resources; their tactical victory was a foregone conclusion.

The statistics seem to speak for themselves. Over three times the tonnage of bombs was dropped in Vietnam as in the Second World War.[6] More armoured vehicles were deployed in South Vietnam alone than the French had deployed in the whole of Indochina.[7] Where the French had eventually been able to call upon a total of 42 helicopters, the Americans deployed about one hundred times as many. Finally, and perhaps most tellingly of all, the Americans seem to have achieved a 'kill ratio' in their battles of between six and ten NVA soldiers to every one American (and by 1972, when the American contribution was largely aerial, a ratio of 34 : 1 was claimed[8]). All this appears to be a conclusive proof that, once they got into their stride, the Americans were incomparably and overwhelmingly superior in tactics. There seems to be no room for doubting that the mobility, firepower and electronic surveillance advantages of a modern mechanised force can destroy a primitive infantry opponent at will.

This argument is persuasive, but on reflection we should perhaps exclude 'electronic surveillance' from the list of important advantages which the American war machine enjoyed in Vietnam. It is true that they deployed many impressive devices in this field during the war, and that they advanced in leaps and bounds as the war continued. The 'McNamara Wall' project to construct a barrier of unmanned sensors along the Demilitarised Zone, for example, stands as a monument to the accelerated research and development of which a rich technological power is capable. It was almost comparable to the 'Manhattan' programme of the Second World War, in its sheer speed of achievement and massive mobilisation of brainpower and industrial capability. In little more than one year an entirely new weapon system was conceived, developed and put into production. The so-called 'electronic battlefield' was converted almost overnight from a cottage craft into a massive industry. It developed its own enthusiasts, its own logistic base, and most important of all, its own bureaucracy and interest-group which could guarantee its continuity. It took its place alongside the 'armour', 'airmobility', and 'artillery' lobbies as an arm of the service. In practice, however, the achievements of the electronic battlefield were as patchy and uncertain as had been those of the tank in the First World War. It suffered from being in its first, experimental generation.

Surveillance was always of paramount importance to the war in Vietnam, since in fighting against a regular army which opted to use guerilla tactics it was always very difficult to find the enemy. Even the best-equipped and most tactically efficient defensive force was powerless against an enemy it could not see. The same problem faced the American main forces in Vietnam which had faced defensive airforces before the advent of radar, or anti-submarine flotillas before the invention of sonar. In all these cases the enemy was able to make undetected raids which could be countered only after the damage had already been done.

There were two main limitations to visibility for the free-world forces in Vietnam. The first was the dense jungle and forest terrain through which the enemy could move almost at will, concealed by his remoteness from free-world bases, and by the canopy overhead. The second was the darkness of the night. It was a maxim in this war, as in the French war in Indochina, that the defenders controlled the daylight hours; but the insurgents controlled the night. Both of these factors imposed enormous frustrations upon the American striking forces, who could venture only cautiously into the jungle, and formed 'Night Defensive Positions' or 'Custer Rings' during the hours of darkness.

To meet the challenge of the jungle, one answer was to physically remove the jungle itself. A proportion of the chemical defoliation operations had this objective, while many experiments were carried out on the ground to find a mechanical technique which would be neither as

141

controversial nor as complicated. In the event neither fire nor gigantic instruments such as the 'Le Tourneau Tree-crusher' (or 'Transphibian Tactical Crusher') were found to be as satisfactory as a simple bulldozer with a specially shaped and reinforced blade known as the 'Rome Plow'. The army found the Rome Plow to be a convenient method of clearing jungle even though its operation was costly in both casualties and mechanical maintenance. Six Rome Plow companies were eventually deployed, and they made extensive clearances along roads and in enemy base areas such as Warzones 'C' and 'D' north of Saigon.[9] Their cuts afforded greater route security and wider fields of fire, although cynics commented on the speed with which the jungle grew again, and on the fact that a cleared road deprived a convoy of its own cover as much as the enemy's. By and large land-clearing did offer a genuine answer to the problem of jungle visibility; but it was slow and confined to relatively small areas.

A second solution to the problem of visibility seemed to lie in the use of aerial surveillance. At the start of the war great hopes were placed in the OV-1 Mohawk surveillance aircraft, a relatively quiet fixed-wing machine which carried advanced surveillance equipment including side-looking radar and infra-red scanners. These aircraft could patrol wide areas in a night, and their early use seemed to promise great things. They suffered, however, from several problems.

The Mohawk's mechanical reliability was often less than perfect especially since payloads tended to be increased above the designed limits. The processing speed of information gathered by the aircraft also turned out to be rather longer than originally hoped. This meant that its immediate tactical 'responsiveness' was impaired, and it came to be regarded as more a strategic than a tactical asset. The commander of the First Cavalry Division (Airmobile) felt that the Mohawk should have been deployed at Corps level instead of absorbing Divisional resources.[10] It was retained at Divisional level, perhaps, because for institutional reasons the army was very anxious to demonstrate its organic need for this aircraft, in the face of strong airforce criticism. The Mohawk was for a long time an important bone of contention between the two services, and this to some extent distorted its practical employment.

Another aerial surveillance platform which was deployed early in Vietnam was the 'flareship' or 'fixed wing gunship' – a cargo aircraft fitted with flare dispensers and electrical Gatling guns. These aircraft were in relatively short supply, but they were highly successful and underwent many improvements as the war progressed. By 1972 a C-130 variant was in service with a 105mm howitzer mounting, as well as auxiliary armament and numerous surveillance aids. Their employment, however, was less for reconnaissance than for supporting fire. They would typically be called upon to provide artificial daylight and suppressive fire around the perimeter of a unit beleaguered at night. They were therefore used to assist in the ground battle once it had started, rather than to find the enemy in the first instance. Over jungle, moreover, their vision was very limited indeed.[11]

Perhaps the most effective aerial reconnaissance was provided by the light helicopter scouting teams from the so-called 'air-cavalry'. These relied at first entirely upon the naked eye, so their usefulness was restricted to relatively clear terrain and daylight. As the war continued, however, their very high tactical flexibility was increasingly appreciated, and attempts were made to extend their vision into the hours of darkness. Especially in the Ninth Division in the delta, but also elsewhere, experimental searchlight attachments (both white light and infra-red) were rigged and image-intensifiers carried. These fittings helped to remove the cloak of darkness from the enemy, while a chemical 'people-sniffing' device could in some circumstances detect a human presence through cover. It was unreliable, however, and fairly ineffective over heavy jungle or in wet weather.

The use of these surveillance aids in light helicopter teams did give a commander additional means of seeing into the night; but they were by no means a complete answer and were not deployed in every operation. They had been unavailable during the early days of the war, and came gradually to the fore only around 1967–8. It is therefore misleading to think of them as a really powerful influence upon the course of the war as a whole, particularly over jungle terrain.

1967 was also the year of the great crash programme to develop unmanned sensors for the 'McNamara Wall'. These were to be dropped by aircraft into the jungle, from where they would send back radio messages if anyone moved near them. Artillery fire or air bombardment could then be placed upon the intruders, or ground forces sent out to intercept them. This system had the great advantage that it at last offered a method of penetrating below the jungle canopy, and could 'watch' wide areas without the need for a garrison permanently on the ground. After three years of war in which the Americans had often been unable to find their enemy, the unmanned sensor seemed to promise a tactical revolution.

The original scheme had been to deploy the McNamara Wall along the whole length of the Demilitarised Zone; but this plan was changed as the siege developed around the Khe Sanh Combat Base in early 1968. The sensors were hastily redeployed around the base and enemy movements beyond the garrison's visibility started to be monitored. The NVA had about two divisions lurking in the area, and the sensors made a contribution to locating them precisely enough for bombardments to be brought down upon their heads. In this role the sensors had the effect of extending the lethal range of the defenders' guns: the cleared *glacis* around a fortification no longer stopped at the jungle edge.

The early sensors at Khe Sanh suffered from many operational defects, although greater reliability was achieved in subsequent models. Many different sensors gradually became available and were deployed in other areas, particularly for the strategic interdiction of the Ho Chi Minh trail through Laos and Cambodia (the 'Igloo White' programme). In a more strictly tactical role they continued to be used as an auxiliary in the defence of firebases, as they had been at Khe Sanh, and also for 'mechanical ambushes' along trails thought to be used by the enemy. Of particular note was the line of such traps laid out and maintained in 1970 around Loc Ninh by the 11th Armoured Cavalry Regiment. This made an apparently very effective contribution to the interdiction of enemy supplies, although it is worth noting that the mechanical ambushes themselves needed fairly constant patrolling on the ground, in order to check results.[12] They were not a complete substitute for ground forces, but really only an auxiliary.

Enthusiasts of the unmanned sensor delight in recounting the tale of FSB Crook, a firebase constructed deep in enemy territory with the specific intention of provoking attack. The base was ringed with sensors and well supported by artillery and aircraft. Everyone hoped that the unfindable enemy would oblige by attacking the base; and sure enough he did so between 5th and 7th June 1969. Early electronic warning helped the defenders to massacre about 400 enemy soldiers for the loss of only one American.[13] The whole idea of battlefield sensors appeared to have been triumphantly vindicated.

The case of FSB Crook, however, does not prove a very great deal. Firebases without sensors had on occasion notched up equally amazing kill-rates, while the role of sensors in firebase defence elsewhere was usually much less spectacular than at Crook. In some cases sensors did not even prevent the enemy from overrunning a position. We should not therefore be dazzled by the achievement of this new weapon in Vietnam, but see it as just one more subsidiary aid.

The vast array of sophisticated American surveillance equipment never provided a complete answer to the problem of visibility in Vietnam. Admittedly it greatly helped the security of firebase perimeters, and eventually turned night into day at will in areas which were relatively clear of vegetation. The jungle nevertheless retained much of its mystery. The land clearing and unmanned sensor programme came too late and on too modest a scale to provide anything but rather local answers to that problem.

If technology failed to deliver the total solution which it seemed to promise in the field of surveillance, surely we can point to its successes in the field of firepower? The fact that firebases containing no more than one or two companies habitually beat off full NVA regiments is a sobering reminder of what modern weapons can do. An attacker could be engaged from within the base by rifles, machine guns, claymore mines, grenade launchers, recoilless rifles, mortars, and artillery firing high explosive, white phosphorous, flechette or cluster-bomb rounds. After short delays additional artillery could be called from other firebases within range, and helicopter gunships might arrive with their rockets and mini-guns. Finally there might be tactical air strikes and perhaps a visit from a flareship. It was a determined attacker indeed who would press forward in the face of this devastating hail of shot. His only chances of penetrating the perimeter lay in achieving total surprise, a silent sapper infiltration, or perhaps a series of improbably lucky hits on key leaders of the defence. The fact of the matter was that American firebases were rarely penetrated and hardly ever overrun. To that extent their firepower can certainly be called overwhelming.

A firebase, however, was a fixed point of known location. Its defenders were dug in and its artillery pre-registered. Heavy pieces of equipment could be emplaced, and open fields of fire cleared all around. It was a fortress in the classic sense, and indeed often had a groundplan which would have been familiar to the great masters of nineteenth century military engineering. Viewed in this light, it need scarcely surprise us that the 'kill ratio' in firebase battles was usually heavily in favour of the defenders: such has always been the case when fortresses are stormed by infantry who are weak in artillery.

If we are to evaluate American firepower in Vietnam we must really ask how well it could be applied in mobile warfare, against targets which appeared at unexpected times and places. This is when firepower was most important to the Americans, since in this type of combat the NVA would hold the advantages of surprise, camouflage and entrenchment. Mobile war was also the most important type of action in the strategic sense, since it was only by taking the war to the NVA that free-world forces could win the initiative. If they had done no more than wait in their firebases until the NVA chose to massacre themselves against the defensive firepower, the Americans would never have been able to impose their own tempo on the war. It would always have been the enemy who chose the level of attrition at any particular moment. It was with this consideration in mind that General Westmoreland adopted 'Search and Destroy' (or 'Find, fix and destroy') tactics in late 1965.[14] It was essential to his strategy that the enemy should be deprived of the initiative by a war of movement.

The problem in mobile warfare was that the enemy was hard to locate before he sprang an ambush. We have already mentioned some of the difficulties of finding him with surveillance equipment. An alternative was to use firepower itself, in a so-called 'reconnaissance by fire'. Thus a unit moving forward into uncharted terrain might call down artillery concentrations in front of it to flush out any NVA who might be lurking there. Preparatory fire was also often used before helicopters touched down in a landing zone which had not been scouted on foot. These speculative fires were sometimes productive and certainly gave confidence to the troops; but they also served to warn the enemy of what was happening. They saved him the trouble of doing his own reconnaissance, and in effect told him when to camouflage himself and take cover. To this extent they operated in a sense diametrically opposite to the one for which they were designed.[15]

It was the fate of the Americans in Vietnam to be ambushed. The enemy usually fired the first shot of an engagement, and in ambushes it is usually the first shot which is the most effective. Only gradually would the ambushed unit be able to patch together its return fire, during which time the ambushers would be able to bring down a prepared barrage from well-sited weapons. The NVA had generally good fire discipline, and were practised in firing volleys at a given signal. At the start of a firefight it was often they, rather than the Americans, who had the upper hand. It was not uncommon to hear of actions in which 'The North Vietnamese had numerical and fire superiority. Initially, it was they who freely employed supporting arms'[16] including mortars and recoil-less rifles.

Ambushed American units had to 'build up a base of fire' piece by piece, if they were to hold their own. They would first reply with their M16 rifles and M79 grenade launchers and a little later these would be joined by M60 machine-guns, which were slightly more difficult to bring into action. In these weapons the Americans enjoyed little superiority over their NVA counterparts, and indeed many Americans who had been shot at with the communist AK47 assault rifle quite naturally imagined it to be more dangerous than their own M16, which had a reputation for jamming. In the case of the M79 there was certainly a clear advantage over the NVA's hand thrown, unreliable and sometimes home-made grenades, although in practice the NVA seemed to get hold of a fair number of M79's themselves. The NVA also used B40 rockets in a similar role, although in this case they lacked the M79's canister round for direct fire at very short range.

The average engagement range in Vietnam was between ten and thirty metres, as a result of the thick vegetation. Even then the two sides would by and large be invisible to each other, and so almost all fire would be unaimed. Since it was fatal to stand up in those conditions, rifles would be fired from ground level and their shots would probably fly high. Only an ambusher who had placed tree snipers overhead, or prepared fire lanes in advance, would be in a position to score many hits after the first volley.

The next American firepower elements to come into action would be the supporting mortars and then the artillery. These had to be called up by radio from the ambushed unit, and this could not always be done quickly. Radio reception might be upset by the lie of the land or the jungle itself. The person calling for support might not be able to see exactly where it was needed, since his area of visibility would inevitably be restricted by his being in a prone posture in thick jungle. In one battle it took a radio conversation of 20 minutes before a unit's grid reference could be established, and even then it turned out to be totally incorrect. There were other cases where local commanders remained unaware that artillery support was required at all, since there was no one with a radio near enough to the point of contact to appreciate the seriousness of the battle, and the jungle tended to deaden the noise of firing.[17]

Once artillery support had been called there would be further delays for registration and clearance. The latter depended not only upon the location of the unit calling the fire, but also on that of any other unit in the area. If the gunners could not establish the whereabouts of every friendly unit it was obviously very dangerous to fire at all. In some cases there might also be restrictions imposed by the rules of engagement. In or near villages or in the Cambodian border area, for example, fire had to be cleared and cross checked for both safety and political acceptability. All this served to delay the arrival of the first accurate shells, and to give the enemy a respite.

Perhaps the most important restriction of all in the use of artillery and mortars was the need to observe safety distances from friendly troops. These might be 100 metres or more, which naturally made the fire somewhat irrelevant to a battle being fought at ranges of 10–30 metres. Enemy formations which were in contact could thus be hit only on their rearward parts or on extended flanks. That could be useful for counter-battery work or for pinning the enemy in place; but his front line fighting portion would often be too close to the Americans to be hit.

Mortars also posed certain problems of their own, since they would often have to be deployed during an infantry operation itself, very close to the enemy. They could not be positioned in jungle with overhead cover, but would have to find a clearing. In practice it was found that the light 60mm mortars used by the NVA were very much more useful than the less manoeuvrable 81mm weapons of the US army. In this particular arm, indeed, it often almost seemed as though the enemy had a distinct advantage.

The effectiveness of mortars and artillery against targets in forests was also reduced by shells striking treetops and exploding before they reached their intended targets, or even ricocheting into friendly troops. It was found, in fact, that only 155mm or bigger guns were fully effective against jungle targets. This immediately downgraded the usefulness of the 105mm pieces which composed about half of the artillery park.[18]

Similar considerations also applied to rocket firing helicopter gunships. Of one battle in the central highlands in 1966 S. L. A. Marshall wrote:

Most of the rockets are HE rounds. Against jungle the ARA gets a lot of detonating that is high in the trees. That scares the people below. But it very seldom hurts anybody.[19]

In another battle the same year Marshall recounts a different type of disappointment with the weapon: 'There were only two gunships. They made a couple of passes, dropped half a load and dropped it in the wrong place.'[20]

This last quotation illustrates two more difficulties with armed helicopters; reaction times and accuracy. Helicopters needed about ten hours' maintenance for every hour's flying, so at any one moment only a fraction of their total numbers would be ready to go into action. Time and again we hear of units with six or eight ships having only one or two 'hot'.[21] Usually more could be readied within an hour or two; but in combat an hour seems to be rather a long time.

A particular difficulty for helicopters was the problem of finding the target. Especially in jungle it was difficult for ground units to mark their own position clearly for an airborne observer, since smoke bombs fired on the ground would not rise above the canopy. In some other cases when marker smoke was fired, the enemy would immediately follow suit with smoke of the same colours to confuse the pilots.[22] It was eventually found that efficient target indication required experienced air liaison officers, either on foot or orbiting overhead, who could keep a proper track of the positions on the ground, and 'talk in' helicopter or fixed-wing strikes. In jungle conditions this was very much an acquired art rather than a precise science. It carried no certainty.

Helicopters could sometimes bring their fire very close indeed to a friendly perimeter, but as with the other supporting arms this practice was risky. The closer the enemy hugged the friendly perimeter, the more reluctant commanders would be to bring down their full firepower. Accidents were all too easy with rockets, bombs and artillery alike. In normal times, at least, it was better to keep these weapons at a healthy distance, and fire well behind the enemy front line. A possible alternative was to back away with your own troops just before the strike went in. In this way you could avoid the risk of being hit by your own supporting arms; but the very act of moving backwards exposed your men to greater risks from the enemy. It was an unenviable choice.

The most awesome weapon in the array of supporting arms was the fixed wing bomber. The weight of concentrated ordnance it could deliver was prodigious, and the suddenness of its intervention was numbing. For a bigger effect, however, a bigger price had to be paid. The reaction time was longer, the safety range greater, and the target indication problem more acute. Even more than artillery and helicopters, the bomber was best used against targets at a distance from one's own front line. It was for interdiction or counter-battery work rather than for the point of contact itself. Its effective use required a good fix on a relatively concentrated target, and we have seen that this was harder to achieve than one might imagine.

In considering the effectiveness of American firepower we must remember that in many cases the target would be dug in. NVA soldiers were expert diggers, and would try to make trenches and bunkers wherever they stopped. When they lay in ambush they would often have only their heads showing above ground, preferably with overhead cover as well. This made them a very difficult proposition to kill. Even if they could be accurately located it took a more or less direct hit from something heavy to knock them out.

The sort of firepower really required by the Americans in their mobile battles was direct fire artillery, or in other words a gun which was much heavier than an infantry weapon, but with sufficient protection to be deployed in the very front line of the fighting. It had to fire accurately outwards from the ground rather than inaccurately downwards from the sky. There was such a weapon in Vietnam, in the shape of the tank; but for a variety of reasons tanks were not available in sufficient numbers in the right places.

The initial appreciation in 1965 had been that tanks were inappropriate for counter-guerilla operations, and that they would in any case be unable to manoeuvre through much of the Vietnamese countryside. They required what was considered at the time to be disproportionately extensive back-up facilities, and their deployment was deemed to be less urgent than that of infantry and helicopter resources. They were slow to appear in numbers, and it was only in 1967 that the armour lobby was able to demonstrate their value conclusively in the detailed MACOV report.[23] By this time it had become apparent both that there was more to the Vietnam conflict than simply the village war against local guerilas, and also that far more of the terrain was practicable for tanks than had at first been supposed.

It was found that a great deal of the jungle could in fact be negotiated by tanks, and that even more of it was passable by armoured personnel carriers. Especially in the warzones north of Saigon this allowed mounted action where previously the war had been conducted mainly by relatively light airmobile forces. In the steep highland regions of the centre and north, however, there were greater problems. In these areas airmobile operations continued to be the most convenient, and tanks were confined to roads and tracks to a greater extent. The tank was thus restricted in its usefulness in many of the most hotly contested battles of the war.

Another problem with the use of armour in Vietnam was one of doctrine. Many commanders preferred to use it for static or road security tasks rather than to send it out in mobile sweeps. This led to excessive dispersion and a failure to realise its true potential. The lesson was gradually learned that armour could be extremely useful in encounter battles; but as with so many other lessons this realisation dawned just at the moment when American forces were starting to withdraw. The solution to the problem came too late in the day.

All the above considerations meant that American firepower was often inadequate to destroy an enemy in contact. Firefights could not be quickly resolved, and they dragged on for hours and sometimes days. It is perfectly true that heavy casualties could be inflicted by artillery or airpower upon an enemy discovered out of contact or in the open; but this did not directly solve the problem of the firefight itself.

For their part the NVA was also capable of inflicting casualties upon Americans out of contact, by the use of stand-off weapons such as mortars, rockets and especially mines. The extent of the mine threat varied greatly from one part of Vietnam to another; but it was always present. A wide range of booby traps, claymores and jumping mines could hit men on foot, while grenades strung in trees could hit men riding on vehicles. For the vehicles themselves there were anti-tank mines and converted bombs or artillery rounds. Against helicopters there were sometimes windmill mines placed in likely landing zones. No complete answer to mines was ever found, and they took a heavy toll throughout the war.[24] This toll was certainly not as great as that inflicted upon the NVA by American stand-off weapons; but it was a constant reminder that technology could not find an answer to every threat.

If American surveillance technology and firepower technology were ultimately inadequate to meet all the challenges of the mainforce battle, what of that other much-vaunted advantage: mobility? The Americans prided themselves on their capacity to multiply numbers by mobility, as in the celebrated case of the marine platoon which made combat assaults in three separate provinces during the course of a single day. Surely this at least was a potent advantage against an infantry enemy who had no option but to walk into battle through thick jungle terrain?

The Americans were able to choose between movement on foot, in ground vehicles, helicopters and (at least by the end of 1966) in a specially designed riverine flotilla. They could parachute in and get out on rope ladders dangling from the sky. On one occasion in the Mekong delta they even experimented with hovercraft. They can certainly be forgiven for imagining that technology gave them an almost limitless range of options.

By far the major part of this technological vision was provided by the new generation of turbine helicopters, which represented an important advance over previous models in range, payload and serviceability. Vietnam was their first major test, and great hopes were placed upon them by their enthusiasts. They were deployed in many varieties and great numbers, and they seemed to open the way to an altogether new style of combat. As one commentator remarked: 'Every infantry unit in Vietnam was, in fact if not in name, airmobile infantry; and its direct support artillery was airmobile artillery.'[25]

In 1965 it was already conventional wisdom that any division of ground troops should have a helicopter force available for liaison, casualty evacuation, command and reconnaissance tasks. What was new was the idea of using large forces of helicopter-mobile infantry in battle itself. They would arrive at their fighting ground in convoys of transport helicopters escorted by helicopter gunships. The gunships would 'shoot them in' to their landing zones, and the infantry would then deploy on foot. For fire support they would be able to call upon still more gunships and massed batteries of 'Aerial Rocket Artillery', as well as conventional artillery lifted in by transport helicopters. Their logistic lifeline would also be supplied by helicopter, and their casualties and damaged helicopters would be lifted out in the same way. The number of lorries and armoured vehicles in the airmobile division could thus be drastically cut, so that the entire formation would be released from a purely overland line of communication. Major parts of it would be free to skip within minutes from one point on the ground to another many miles distant. They could keep the enemy guessing and off balance.

This new idea had already been tested on a limited scale both in the USA and in Vietnam; but it was in 1965 with the deployment of the First Cavalry (Airmobile) Division that it really saw its full embodiment. It was then that the army's helicopter enthusiasts were given the opportunity to demonstrate just how much they could achieve. In particular, they felt they had to prove that it was only by airpower under organic army command that the best results could be obtained. They were very concerned to undermine the airforce claim that all tactical aircraft should be an airforce responsibility. It was partly for this institutional reason, in other words, that the army committed itself so massively and so suddenly to the helicopter in Indochina.

The helicopter idea certainly had much to recommend it, and it is extremely doubtful that the war could have been fought at all without it. On the other hand it also proved to have several important defects. The most significant point, perhaps, was that helicopter troops suffered from exactly the same problem which bedevils all other types of airborne or airmobile troops – they have to be extremely careful about where they choose to land. They cannot set down in terrain which is either too rough in itself, or controlled by enemy fire. In Vietnam this meant that although helicopters might be able to move rapidly from one landing zone to another, their choice of landing zones turned out to be relatively limited. One commentator has remarked that 'in fact the Americans simply became as tied to their helicopters as the French had been to their roads'.[26] There is a great deal of truth in that statement.

The helicopter was an amazingly versatile and flexible weapon in Vietnam; but it could not land in thick jungle. It had to find a natural clearing, or perhaps have an artificial one blown or laboriously hacked out for it. As the war went on several different techniques were also developed for depositing men into the jungle from a hovering helicopter: but such methods could never provide a large-scale substitute for the ability to touch down at will wherever the helicopters were required. Casualty evacuation and resupply both tended to demand a properly cleared landing zone; and these two functions were both essential to troops in combat for any length of time. The result was that many operations were limited to areas which had landing zones nearby. The foot mobility of the fighting men was significantly hampered by the need to maintain this umbilical cord.

The security of landing zones was also important, since it was highly risky to land a helicopter under close range enemy fire. Airmobile soldiers always had a deep-rooted horror of running into a landing zone which was covered by snipers – or as Philip Caputo put it, 'Happiness is a cold LZ'.[27] As the war went on the NVA became increasingly effective in shooting down helicopters, as both anti-aircraft weapons and ambush skills improved. Particularly during the 1971 incursion into Laos they were able to stake out every jungle clearing which could possibly be used as a landing zone. Almost every helicopter assault met an ambush, and a total of 107 helicopters were lost in the operation as a whole.[28] This total was untypically high, perhaps, because bad weather greatly restricted the times and altitudes which could be flown. It is nevertheless a striking illustration of the need for secure landing zones.

Landing zones were secured by infantry on the ground, who might arrive either on foot or in helicopters. In either case they might initially have to fight their way in against opposition; but their task was essentially a static defensive one. It was expensive in the one resource in short supply – manpower. American battalions in Vietnam usually went into action well understrength, so every detachment for landing zone security bit deep into total offensive fighting power. This task perhaps did as much as anything else to blunt the effectiveness of mobile operations.

Troops who were actually in contact, furthermore, were often deprived of helicopter support at the very time they needed it most. It was highly dangerous for resupply, reinforcement or medical helicopters to touch down in the middle of a firefight. They were forced either to wait until the firing had subsided, or to find an alternative landing zone at a distance. In either case the help they could bring to the ground troops would arrive late. It is true that in these circumstances many brave pilots ignored the risks and flew straight into the thick of the fighting; but that was not a recommended practice. It was the most certain way to lose helicopters.

The problems of selecting a landing zone were serious indeed; but airmobile infantry also suffered from a second handicap common to all airborne troops. They were relatively lightly equipped, and in most operations their ground movement was on foot. It is true that their supporting artillery batteries moved regularly by helicopter, and in some cases armoured vehicles did so too. On one memorable occasion even armoured boats were lifted bodily out of the water and set down in a new tactical area. In general, however, airmobile soldiers were transformed into foot-mobile soldiers as soon as they arrived on the ground.[29]

This meant that when they met the enemy they had little advantage in the firefight apart from their supporting arms called from a distance. When ambushed, an American unit was quite likely to be pinned down by NVA fire. Its mobility was suddenly limited to the speed with which an infantryman could crawl through thick vegetation: no more than

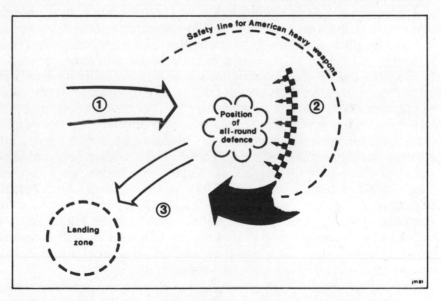

Schematic view of a 'typical' mobile sweep.

1 *Americans advance to contact.*
2 *Americans pinned down by NVA ambush fire.*
3 *NVA manoeuvre to surround Americans. Americans try to find a clear landing zone for casualty evacuation and re-supply.*

about one metre per minute at the outside. Instead of 'finding, fixing and destroying' the enemy, it often became a case of 'finding, being fixed and fighting a desperate battle to regain freedom of action'.

When American infantry found themselves walking into close range fire from a dug-in enemy, the only possible response was to hug the ground until the incoming fire abated. A unit's movement was halted by bullets even more than by the jungle itself, and of course this was even more true in open ground: 'Once a person gets hit, and your fire and manoeuvre stops in a paddy, your momentum is dead. It gives the enemy a chance to sight in. When the next man gets up, he'll get dinged – then nobody wants to get up.'[30] This was a most important limitation upon American mobility in Vietnam. As one NCO put it: 'Every time I manoeuvre a man, I get him shot, and I say to hell with it.'[31]

NVA firepower thus presented a genuine physical obstacle once a firefight was joined. Units simply did not advance as they were supposed to. This was compounded, however, by the problems of casualty evacuation. It was a point of both policy and honour to extract all killed and wounded soldiers as fast as possible. This act was generally considered to be 'more sacred than life itself,' and it frequently distorted the shape of entire battles.[32]

The evacuation of casualties under fire is at the best of times a tricky operation. It involves great courage on the part of those who venture forward into places known to be beaten by enemy fire, and it can easily lead to their becoming casualties as well. It distracts all the other soldiers in the immediate vicinity, several of whom will probably have to break off the battle to carry each casualty to the rear. Perhaps most important of all, it distracts the local commander by making him responsible for arranging transport out of the battle area. Not only will his own mind be diverted from the battle itself, but his reports may implant misleading impressions in the minds of higher commanders. If they hear nothing of the fight apart from the details of casualty evacuation, they can very easily be led to assume the worst. They may imagine that each casualty evacuated represents only the tip of an iceberg, and they may lose perspective. In Vietnam radio communications allowed higher echelons to hear all these details as soon as they were reported, as well as to make hasty changes of plan as a result. Good radio communications did not always serve to clarify the position on the ground.

In one firefight, for example, a platoon commander received a hit in the eye, yet continued to report the battle on the radio. He did not reveal the fact that he had been wounded, but suggested instead that the evacuation of other wounded men was creating a crisis: 'These bastards [i.e. the NVA] have too many MGs, and now the SOB's are rolling us with grenades. The platoon is becoming split, due to care-for-the-wounded pulling the able bodied in opposite directions.'[33] When he came off the air the picture he had painted was not corrected, and his superiors felt they had to send direct relief. In the event, however, the casualties had not been nearly as heavy as reported, and the relief had not been required.

There are many examples of operations which were virtually halted in their tracks as soon as it became necessary to evacuate casualties. We hear of one company commander who, at the height of a battle, 'was now worrying more about getting a medevac ship in than about hunting the enemy'.[34] In another case we hear of a South Korean unit, whose soldiers 'were interested in getting their own back, whether the quick or the dead. After that they would deal with the enemy.'[35] Care for the wounded often seemed to be given a higher priority than the continuation of battle.

This humanitarian reaction doubtless originated among the 'buddies' of men who had been hit.[36] Their care for each other in danger is a striking illustration of unit cohesion, if not of aggressive instincts. On the other hand immediate casualty evacuation was also called for by higher command echelons and had both a strategic and a political rationale. We must remember that Westmoreland's strategy was, *faute de mieux*, one of attrition. He had to inflict very many more casualties upon the enemy than they could inflict in return. It was therefore an important part of American strategy to lose as few men as possible. In order to keep the American public interested in the war, furthermore, the government wanted to demonstrate that it was doing everything it could to ease the lot of the combat soldier. The provision of an elaborate medical apparatus was thus of relevance to both these concerns.

It was felt that there was something contradictory about being killed in a limited war. If there was any chance at all of saving a wounded man, everything possible would be done for him. The statistics show that in this endeavour the Americans were astonishingly successful. A much smaller percentage of casualties died of their wounds than in any previous war. Within minutes of boarding the medical helicopter a combat victim could be receiving expert care, or even be en route to a base hospital in Japan.

It was not only the wounded who were evacuated. The killed were also shipped out of the battle area with rather more haste than in any previous war. In this case it was not so much the 'limitation' of the war which provided the rationale; but its psychic barbarity. There was a feeling that Vietnam was somehow not a civilised place to leave the bodies of American dead. Soldiers in action would sometimes expose themselves to enormous risks, and even death itself, in order to retrieve their comrades who had been killed. The same, it must be admitted, was also true of the Vietnamese soldiers on both sides.

In this very specialised sense, at least, we can call the Vietnam war one of the most humane there has ever been. The Americans often stopped their attacks in order to look after their casualties. Combined with the difficulties of the terrain and the volume of enemy fire, this consideration seriously impaired mobility on the battlefield. Colonel D. R. Palmer has gone so far as to suggest that it was the most important consideration of all. In a passage in some ways reminiscent of late nineteenth century French writing about 'morale', he claims that mobility is a state of mind rather than a purely physical attribute.[37] He suggests that American infantry attacks could actually have maintained their momentum if both commanders and men had been less anxious to minimise losses.

One indicator of attitudes in any army is its training doctrine, and we find that the American doctrine for infantry combat changed radically in the first few years of the Vietnam war. At the start it had been theoretically accepted that infantry could fight its own way forward into enemy positions, using 'fire and manoeuvre'. A unit which made contact would first build up a base of fire to pin the enemy to his positions. Under cover of this fire the infantry could manoeuvre forward in small groups and skirmish into the enemy area. There would then be a brief phase of combat at close quarters which, although it might be costly in friendly casualties, would lead to the annihilation of the opposing force. In 1965 both the Army and the Marine Corps had recommended a form of this procedure in their training manuals.[38]

In practice, however, 'fire and manoeuvre' was rarely used in Vietnam. There turned out to be both a carrot and a stick which tended to lead commanders in a rather different direction. The carrot was the unprecedented weight of supporting firepower which was now available to even the smallest unit; while the stick was the political requirement to avoid casualties. Taken together, these two factors inevitably encouraged the idea of substituting heavy supporting fire for the costly assault phase of an attack.

It came to be believed that a base of fire could be built up with sufficient power to physically destroy the enemy, rather than merely to pin him down. In this way the decisive action of the battle could be entrusted entirely to supporting arms, so that a final infantry assault would become superfluous. The infantry's role would be limited to finding the enemy in the first place, and then helping to bring down fire upon him. As Colonel Palmer has remarked, this amounted to a tactic of 'Manoeuvre and Fire'.[39] The infantry found itself excused by technology from the dangerous need to stand up and advance after battle had been joined. In 1967 these tactics were codified in a training manual written by Colonel S. B. Berry and quickly given official blessing. In the conditions of Vietnam there was perhaps only one way in which American infantry could have been expected to manoeuvre under fire; and that was by riding in armoured vehicles. Inside a tank or an armoured carrier the soldiers enjoyed a relative immunity from fire, and their mobility was restored. On many occasions it was found to be perfectly practicable for these vehicles to charge forward into the midst of enemy positions, sowing mayhem and confusion as they went. It was an effective technique; but it was not one which lent itself to airmobile infantry. They had no armour heavier than their cumbersome flak jackets, and in effect this left them with no alternative but to adopt the new tactics of 'manoeuvre and fire'.

The significance of the new tactics is that they assigned a purely passive or defensive role to infantry in the firefight. When they made contact, American troops almost always adopted a position of all-round defence, firing outwards and waiting for their supporting arms to become effective. As S. L. A. Marshall put it, 'there had been no hand-to-hand fighting in the literal sense. Every firefight had had the nature of a small siege with the Americans holding the ramparts.'[40] Even in supposedly mobile operations the tendency was to set up improvised perimeters which soon began to look very much like minor fire support bases. The onus of manoeuvre was handed to the enemy.

In some cases the NVA accepted the challenge and made frontal attacks upon the immobile American infantry. These assaults were generally expensive and rarely successful. More often, however, the NVA would try to move round a flank and complete an encirclement. In many ambushes the soldiers in front of the Americans would remain dug in and pin them with fire, while a mobile force made a wide sweep and started to infiltrate the position from the rear. This attempt was by no means always successful; but it did serve to increase the pressure on the beleaguered American infantry. It kept the initiative in communist hands.

In cases where complete encirclement was successfully achieved, the strains on the Americans could become very great indeed; both for those inside the threatened circle and for higher commanders outside. The whole character of an operation could be changed in an unexpected and unwelcome way. As Marshall put it, 'More than any other distraction in war, the unit cut off and fighting for survival is likely to make a battle plan fall apart.'[41] Or again:

It is a situation that too frequently occurs in the Vietnam fighting. The forward element, losing men and becoming pinned down, compromises the position of all the others. What has started out as an attack loses all form and deteriorates into a costly rescue act.[42]

Units of any size, from the biggest to the smallest, could and did become enmeshed in battles of this type. At Khe Sanh Combat Base in 1968, for example, it was a reinforced regimental group which found itself surrounded. Its predicament distracted the attention of the White House for the space of several entire weeks, and possibly shifted the whole balance of the war. At the other end of the scale lies the saga of 'BAT 21', an EB-66 aircraft which was shot down in 1972 in the Demilitarised Zone. There was known to be one surviving crew member; but he was surrounded by the best part of an NVA division. A battle lasting twelve days was fought to extract him, during which even B-52 bombers were used, and several more aircraft and helicopters were lost. The attempt was ultimately successful and the enemy was severely mauled into the bargain; but as an example of a rescue distorting the shape of other operations it must stand unsurpassed.[43]

There is no doubt that the Americans became highly proficient in extracting both surrounded units and individual soldiers from tight corners. In so doing they were often also able to inflict very damaging losses upon the enemy. This should not, however, blind us to the fact that in each one of these battles it was the NVA who really held the initiative. It was they who manoeuvred to surround the Americans, and they who chose when to make a fight of it. At the point of contact on the ground, where it mattered most, it was they who continued to enjoy superior mobility.

The NVA also enjoyed superior mobility in another respect, which was in some ways even more important. This was that they could usually break off combat when they chose. The American hope that supporting arms could destroy their opponents without the need for an infantry assault was rarely realised in practice. In firefight after firefight it transpired that the NVA had slipped away before being genuinely destroyed or overrun. They certainly suffered heavy casualties; but they were prepared to pay this price if it meant that their units could survive as working organisations. The NVA recognised that there was a significant difference between losing a proportion of their soldiers and losing a unit as a whole: they consistently chose the former option.

Exfiltration from the scene of combat was a highly developed art among the North Vietnamese forces. When the order was given to disengage, they could split down quickly into small groups to crawl out of the battle area in every direction, exploiting the terrain and American bafflement to the full. Only much later, at a good distance from the scene of the fighting, would they regroup at a pre-arranged rendezvous and count their losses. This formed an integral part of their plans, and it was extremely difficult to counter.

From the American point of view the fact of enemy disengagement could usually be represented as a success. It removed the pressure from beleaguered infantry and left them in control of the battlefield. In terms of relative attrition also it often appeared to be highly profitable. There was rarely, however, very much that the Americans could do by way of pursuit.[44] After each firefight they would typically be more concerned with casualty evacuation and policing the battlefield than with following up their elusive tormentors. Like Napoleon after Borodino or Bautzen they would find themselves disappointed that the spoils of a hard-fought victory appeared to be so slight. They failed to win any of the spectacular trophies which are inseparable from truly decisive military action.

It was the indecisive nature of their firefights which ultimately became so frustrating to the Americans. Unable or unwilling to overrun enemy positions while they were still manned, they relied upon the firepower of supporting arms to do the job for them. When this assistance turned out to be less than decisive, they would realise belatedly that they had lost the initiative. The enemy was free – admittedly at great cost – to manoeuvre, disengage and fight another day. As Philip Caputo remarked of his first battle: 'Like so many of the thousands of firefights that were to follow, it began with an ambush and ended inconclusively.'[45]

Truly decisive battles seemed to require the complete encirclement of entire enemy units, so that not even small groups of soldiers could worm their way out of the trap. This realisation came home to the Americans very early in the war, after the battle of Ap Bac I in January 1963. On that occasion a South Vietnamese commander had allowed an enemy mainforce battalion to escape from contact after it had inflicted considerable damage upon his greatly superior forces. There was more than a suspicion that he had deliberately given the Viet Cong a 'golden bridge' for their escape, in order to spare himself the costly necessity of making an assault. With this cautionary tale before them, therefore, the Americans came to Vietnam with a determination that the same thing should not happen to them.

There is no doubt that American commanders at all levels had the intention of inserting fresh units behind an enemy to cut off his retreat at the same time as he was being engaged frontally by fire. They looked for battles of annihilation in which encircling or blocking forces would be just as important as those involved in the main firefight itself. There had to be an 'anvil' as well as a 'hammer', a 'seal' as well as a 'jitterbug', or a 'cordon' as well as a 'sweep'. All this was well understood, and the literature is liberally sprinkled with just those phrases. The difficulty in mobile warfare, however, lay precisely in providing the 'anvils', 'seals', or 'cordons' which were required.

We have seen that an American unit which actually made contact was often effectively prevented from manoeuvring or destroying the enemy on its own. It would call for outside assistance while remaining pinned to the spot. This posed a double problem for the higher echelons of command, since they then had to think both about giving immediate relief to the pinned unit and at the same time about inserting other units as blocks behind the enemy. All this often proved to be an excessively complex requirement in the time available, and it usually happened that although the Americans might achieve the former task, they would fail in the latter.

A major difficulty was to be found, as so often in Vietnam, in the nature of the terrain. As Marshall explained, 'The jungle is simply not meant for fighting. It mocks manoeuvre.'[46] It proved to be no easier to lay on a timely intervention on the rear of a firefight which flared unexpectedly than it had been to manoeuvre the units actually in contact. Available reserves had to be scraped together without much notice, and secure landing zones had to be found at the right distance from the fighting. Once the reserves had been safely deposited on the ground they still had to beat their way forward through the underbrush to a point where they would be useful – usually giving plenty of warning to the enemy as they came, and quite possibly running into a new ambush. Even when they were finally in place there was little guarantee that they could effectively cover all avenues of escape.

The NVA was highly skilled in the art of exploiting the slightest gap in a surrounding cordon, and in jungle conditions that gap might only need to be a few feet across. When we add to this the need for the cordoning troops to adopt an all-round defensive position, we can see that the operation required a very large number of men indeed – perhaps one for every metre of the cordon. In many cases the net could only be made complete by using artillery or air bombardment to fill the spaces between infantry units; but the enemy seemed willing enough to accept the losses entailed in passing through such a barrier of fire. Especially at night it was almost impossible to prevent this escape through the jungle. It is therefore with some reason that the Americans are particularly proud of those few battles in which the cordon did work as planned.

To some extent all this was a matter of 'sending a boy to do a man's job'.[47] Because they had excessive confidence in their technological edge, the Americans perhaps felt they could make do with fewer infantry on the ground and fewer reserves than the task actually required. Their planning perhaps took too little account of the local difficulties, and assumed that each infantryman would be able to multiply himself by firepower and mobility rather more than was actually the case. At all events it seems clear that they often committed too few men to each action, with inadequate genuinely effective firepower and mobility. Contrary to the popular belief, the Americans deployed too little of their much-vaunted combat power, in these battles, rather than too much.

As for the enemy, he seems to have struck upon a highly effective way of multiplying manpower and firepower by combat technique. The key seems to have been constant movement and a willingness to accept high casualties when taking the initiative, combined with the use of terrain, camouflage, dispersion and entrenchment to minimise casualties at other times. In this way the NVA was able to keep an effective army in being, even though it was technologically unsophisticated and relatively lightly armed.

When an NVA regiment passed into South Vietnam from its sanctuaries in Laos or Cambodia, it would typically move in small groups to a large pre-prepared and fortified base area situated in difficult terrain. In this base it could re-group, study local conditions and conduct rehearsals for forthcoming operations. While it was in its base area, furthermore, it was well poised to accept battle on favourable terms with such American units as might stray in. In the event of a co-ordinated large-scale sweep it could refuse action and retire in small groups. In normal times, however, it would enjoy almost as great an immunity from attack in its bases as the Americans did in theirs.

A central operational principle for the NVA was known as 'one slow, four quick'. This meant that their attacks or major ambushes would be meticulously planned and prepared well in advance. Arms, food and medical supplies would be sent out ahead of the combat troops and cached near the intended scene of battle. Only when everything was fully ready would the infantry make their four quick actions. First there would be a rapid movement, still in dispersed groups, to the battle area. Then a sudden concentration on the field itself would deliver a violent and unexpected blow at the decisive point, covered by ambush parties on the flanks to confuse and delay enemy relief attempts. The third phase was a quick but thorough policing of the battlefield to collect weapons and casualties. Finally there would be an equally rapid withdrawal to a known rendezvous point.

This technique by no means guaranteed success at the decisive point itself, where a dug-in firebase or a well-protected convoy would often manage to fight its way out of difficulty. The technique did, however, frequently achieve the advantages of surprise and ensure that regrouping was effected with the minimum of consternation and panic. It kept the initiative with the NVA at every stage, and sometimes stretched American mobility and 'responsiveness' to the limit. In important respects it forced the Americans to 'run in order to stand still'. They certainly needed all the technological assistance they could get if they were to avoid a defeat comparable to that of the French before them.

The NVA was a regular army using many of the tactical methods of the guerilla. It proved to be almost a match for the Americans in the mainforce battles, and we may perhaps ask whether its technique might not have a general relevance to the future of warfare. Already in Burma during the Second World War we were hearing of both Chindits and the 'one man front'; while soon after the war a number of theorists were suggesting that dispersion and guerilla methods would have an increasing part to play in the regular operations of the future. In Vietnam there seems to have been no more than a continuation of this existing problem.

Vietnam was also characterised by a patchwork of fire support bases supplied by air. These too had precedents in the Second World War, as most notably at Imphal-Kohima and Bastogne. In these cases there was a move away from a linear defence towards an 'archipelago' defence. It was realised that even a relatively small force could hold out indefinitely if it disposed of air resupply plus the defensive firepower of modern weapons. This, once again, would seem to be a pointer for the future.

The most important tactical lesson of Vietnam, however, is that even the most advanced technology may fail to deliver everything that it seems to promise. The Americans found that for all their superabundance of modern equipment they still lacked a surveillance system capable of finding everything that had to be found. They lacked firepower which could hit everything there was to hit; and they did not have the mobility to go everywhere they wanted to go. This in itself must have amounted to a great disappointment; but it was compounded by a tactical doctrine which often seemed to ignore such embarrassing technical shortfalls. It is almost as if the Americans assumed that their equipment would work to the most optimistic manufacturer's specifications. They seemed to believe that supporting firepower really could destroy an enemy without the need for an infantry assault; that rapid casualty evacuation could be performed without tactical dislocation; and that blocking forces could easily be inserted behind an enemy with speed and effect.

American tactics in the mainforce war were not always fully appropriate to the hard realities of either the jungle battlefield or the available equipment. They led to victories which were less than complete, and a rising sense of frustration with the war. There were many other causes of this frustration, to be sure; but from the combat soldiers' point of view it must surely have been the deficiencies of battlefield technology which were of especial importance.

The Rambo Generation and the War After Vietnam

When 'Forward Into Battle' was first written in 1981, the Vietnam war was still too fresh and too painful for there to be much American commentary, literature or cinema about it. Just as the period from 1918 to 1927 had been something of 'a time of silence' for British and French commentators on the Western Front, so the nine years of shock and national re-assessment immediately after the 1973 'peace' treaty left the Americans strangely mute. Vietnam was quietly excised from West Point curricula, the army made it clear that it had never really liked the draft after all, and decent citizens made vows that they would not raise such a painful subject as Vietnam in public conversation. The very words 'counter-insurgency' and 'counter-revolutionary' easily reverted to their pre-Kennedy level of odium. The American military press today routinely carries articles on 'low intensity' warfare instead.

Also similar to the reaction after 1918, 'The Imperial Republic' soon found that the very idea of overt intervention overseas had became a potent taboo. Vietnam had been the foreign adventure to end all foreign adventures. Just as there were to be 'no more Passchendaeles' in the 1920s and 30s, so there were now to be 'no more Hamburger Hills' in the 1970s and 80s. President Ford won but little kudos by his rearguard 'Mayaguez' operation, but highlighted military fumbling when a costly battle was fought to save hostages who were already safe.[48] President Carter fell from grace with his own abortive hostage rescue in Iran; and President Reagan did no good either by his still more tragic intervention in Beirut. Massive deployments to El Salvador and Nicaragua were ruled out, and the Soviet Union was effectively given *carte blanche* to find its 'place in the sun' in Angola, Ethiopia, and Afghanistan.

By the mid-1980s, however, a sea change had occurred in the Vietnam debate. The desire never to repeat such a war was still as strong as before, but it was now mixed with a flood of books and films which drew the images of Vietnam out into the open, out of the shadows of the repressed subconscious. The 'survivalist' cult flourished, and even mainstream fashion was influenced by jungle fatigues and combat clothing. A wave of rightist revanchism appeared, which sought to ensure that America should walk tall once again, and which perceived as weakness the limitations on action imposed by fighting a limited war. 'Someone wouldn't let us win,' complained Rambo, and in this he expressed the thoughts of a generation.

The new national mood released foreign policy from some of the shackles of its earlier rules of engagement. When Grenada was occupied and Libya bombed, the applause from the domestic audience was frenetic. When Colonel North was arraigned for bypassing the constitutional niceties, he was hailed in many patriotic circles as a hero. In strictly military affairs, also, there was a corresponding injection of new funds and a toughening of attitude. Vietnam now became a humiliation to be avenged, or at least a parable to guide future action. The military accepted a version of what had happened which largely absolved it from direct blame,[49] but which strengthened its resolution never again to accept false strategies from outsiders and businessmen.

Perhaps the most direct impact of Vietnam on military policy, apart from abolition of the draft, was the construction of a new conventional version of the nuclear 'massive retaliation' thinking of the 1950s. In Korea General MacArthur had been baffled by the restrictions of limited war, and had declared that 'there is no substitute for victory'. But because this would have implied the use of nuclear weapons and the start of World War III, he was promptly dismissed by President Truman. In the 1960s, therefore, the military establishment reluctantly accepted all the restrictions on its Indochina war that were placed upon it. These included respect for the neutrality of Cambodia and Laos, and a bombing campaign in North Vietnam that was only 'graduated'. By the 1980s, however, both of these approaches had been discredited.

The air force pointed to Operation Linebacker II at Christmas 1972, when the B52s had finally been released to bomb Hanoi. Not only had this led to a surprisingly light toll of the attacking aircraft, flying against the most strongly defended target in the world, but it seemed to be successful in 'bombing the enemy to the conference table'.[50] It stood in stark contrast to all the futile years since 1965, when countless aircraft had been lost attacking negligible targets, producing no political effect except a strengthening of international hostility. For the future, therefore, the air force was keen for a 'gloves off' policy, from which the first fruits were perhaps seen in Libya in 1986.

The army's analysis of the Vietnam experience concentrated on the question of attrition as opposed to hot pursuit. In the event it had been compelled to fight a purely defensive war, forbidden before 1970 to pursue the invaders to their sanctuaries in Laos and Cambodia. The defensive had necessarily been 'active' and firepower-oriented because the US forces lacked sufficient manpower to hold a continuous fortified line along the South Vietnamese frontier. This led to constant recriminations, since whenever ground had been won in some bitterly-contested firefight, it soon had to be abandoned to allow the troops to go searching and destroying elsewhere. The attritional aims had not been achieved, either, since the North Vietnamese had always been free to remove their forces to safety whenever they wished, and hence to accept casualties only at a level they believed they could bear. In this way they were given time to turn the weapon of attrition around, so that its decisive bite was felt against US servicemen instead of against themselves. It was the steady trickle of returning body bags that finally lost Uncle Sam his war with the American public.

Nor was the statistical basis of attrition a success, with the 'body count' and 'hamlet evaluation scheme' notoriously open to corruption and abuse. There was a widespread feeling that although Pentagon data banks might be brimming with completed questionnaires and 'management indicators', these were an inappropriate tool for generating strategic insights or fighting a real shooting war. In many quarters Secretary of Defense Robert McNamara[51] was held to be personally responsible for foisting the abstract and impersonal methods of business accountancy upon a profession which ought to have looked more to charismatic leadership and traditional brands of wisdom.

This particular debate came to a head in the early 1980s as a result of attempts to assimilate a new generation of weapons for conventional warfare. In its field manual FM100–5, dated 1976, the army had accepted a style of 'active defense' in Germany which relied on moving forces sideways from quiet sectors in the front line, to concentrate their effort and firepower against incoming enemy spearheads at identified points of threat. Fire from aircraft and rockets using newly emergent technology, or 'ET',[52] would provide a further 'combat multiplier' that could allow the defender to fight outnumbered and win. To give this its maximum effect, statistical calculations were developed to estimate the precise amount of fire that would have to be stacked up against each enemy spearhead in order to stop it; or, in the jargon, how much 'servicing' each target would require.

This doctrine for conventional warfare enjoyed the full backing of the arms industry and TRADOC,[53] but it was soon being questioned by those who saw all too many parallels with the approach used in Vietnam. 'Active Defense' was criticised for being inactive and unaggressive, relying once again on attrition, firepower and a 'managerial' fix to a military problem. At an operational level it involved waiting passively for the enemy to choose his own time and place to advance into friendly territory, then it invoked statistical mumbo jumbo to calculate a level of attrition that might stop him. Manoeuvre would be relevant only out of contact with the enemy, as 'reserves' were moved away from sections of the front line which were thought unlikely to be threatened. This in turn would mean that valuable real estate was left abandoned and open to subsequent occupation by the enemy, just as it had been in Vietnam. As in Vietnam, also, firepower was still supposed to do the trick almost alone. Yet the new ET weapons, from which so much was promised, could easily be seen as little more than descendants of the 'electronic battlefield' which had achieved only very mixed results in combat.[54]

As the salesmen of ever more elaborate ET weapons sought to refine their doctrine for 'deep strike' and the air-land battle of the year 2000,[55] many military men recoiled instinctively against their root assumptions. Notable among these was General Starry, the new commander of TRADOC itself, and earlier a champion of armoured mobility in Vietnam.[56] His vision of the 'extended battlefield' included ground forces supplementing standoff firepower strikes with armoured and heli-borne counter-attacks. By hitting the rear or forward enemy echelons, their front-line troops could be starved of logistic support and would be easier to fight in the normal way. By hitting subsequent reinforcement echelons, the enemy already in action could be kept to a manageable number. This became a very influential line of argument indeed. Its more technological strand was taken up in the 'Follow-on Forces Attack' (FOFA) concept elaborated by General Rogers, NATO's Supreme Allied Commander Europe. Meanwhile, a less hardware-based version surfaced in the army's new edition of FM100–5, issued in 1982.[57]

The new field manual was the army's considered response to new technology, but inspired by the lessons of Vietnam. With some distinct echoes of de Grandmaison, it called for a return to the spirit of the offensive, seizure of the operational initiative and a rejection of the all-firepower, all-attrition style of battle. Even though NATO was still forbidden to initiate war by launching a spoiling attack or pre-emptive strike, as the Israelis had done in 1967, at least a number of other restrictions could be lifted. There could be decisive manoeuvre and hot pursuit, such as the Israelis had demonstrated by their canal counter-crossing at 'Chinese Farm' in 1973. If the enemy entered friendly territory, he had to understand that we might not rest purely on the defensive, but might retaliate back into his territory: we might 'take tokens' from him, to pressure him into making peace and relinquishing what he had taken from us.[58] In short, in the crude behaviourist terms so dear to revanchists, the enemy had to be made to pay.

Deep strike, the extended battlefield, FOFA and FM100–5/1982 all shared a willingness to use conventional weapons, not only within sight of NATO's front line, but far beyond the horizon and sometimes hundreds of miles beyond. Not even the most gutsy disciple of Patton wanted to dispense completely with long-range ET weapons, electronic surveillance aids or airmobile assaults, although there was a considerable debate about just how far ahead of the fighting troops one should try to look, shoot or fly helicopters.[59] On the whole the 'soldiers' tended to want plenty of reliable assets manageably close to their own FEBA, while the 'scientists' were often more ready to gamble on fewer but more sophisticated weapons designed to achieve wonderful things at greater distances. What neither of them, nor the burgeoning 'light infantry' lobby, was apparently ready to swallow, however, was a low- to medium-tech, manpower-intensive 'non-provocative defence' such as was being recommended by a number of European interest groups, not uniquely on the far left.[60] It was apparently an irreducible feature of America's military culture, no less than of its geo-political and economic situation, that any large force maintained overseas had to be especially strong on hardware in proportion to its manpower. If large masses of foot soldiers were required, they should be provided by the host nation. This was an approach that had been very evident in Vietnam, but it was not subjected to questioning thereafter. An American army corps in Germany is still almost twice as strong in heavy weapons as some of its West European equivalents, and a 'brigade' can sometimes be almost as strong as the 'division' in some other armies.

As far as the purely tactical lessons of Vietnam are concerned, there is a clear continuity in American thinking between their practice in South East Asia in the 1960s and their projections for a possible battle for Germany in the 1990s. It is slightly ironic that they were criticised for

166

trying to master a low-intensity Viet Cong campaign using high-intensity NATO weapons, since today they have brought updated versions of their counter-insurgency equipment back into the big league of European warfare. The A-10 anti-tank aircraft, for example, is modelled on the slow-moving, heavily-armoured, propeller-driven close-support aircraft of the jungle war, just as the Blackhawk and Apache attack helicopters represent the third generation of development from the humble 'Huey Hog' rocket-firing Iroquois. Furthermore, these weapons would often be used in roles similar to those seen in Vietnam, delivering heavy air-to-ground prep fires to suppress flak, followed by assault from heli-borne light infantry protected by gunships.[61] Whether or not the dense air defence environment to be expected in Europe could actually be suppressed by these means is a moot point. In 1982 the Israelis successfully mastered Syrian defences in the Beka'a Valley, and there has been much development of both ECMs (electronic counter-measures) and 'smart' flak-seeking ammunition even since then. On the other hand, the vulnerability of anything flying low and slow surely remains axiomatic.

The technology for electronic battlefield surveillance has mightily advanced since its early halting steps in 1967–8. The 'Buffalo Hunter' RPVs (remotely piloted vehicles) which turned upside down and photographed the sky instead of the Son Tay POW camp in 1970[62], have been replaced by 'Super-drones' capable of relaying secure realtime television pictures of the evolving combat. AWACS and the TR-1 now supplement the Mohawk, and a brand new range of ground sensors can 'wire the battlefield like a pin-table' far more effectively than was possible on the Ho Chi Minh trail during the Igloo White programme. Night can be made into day as never before, although in fairness it must be remembered that the flareships and infra-red scanners widely deployed in the 1960s had already moved a considerable distance down that road. When the Israelis were caught out without them on the Golan in 1973, they were merely behind the times.

In armoured warfare itself the Americans now have a brand-new set of tanks and APCs, and even the advanced Sheridan missile-firing tank of 1966 is today considered something of an anachronism. The powerful turbine-engined M1 Abrams main battle tank perhaps owes little to the Vietnam war apart from its name; but there is continuity through the Mechanised Infantry Combat Vehicle and Infantry Fighting Vehicle, which incorporate lessons learned with the ARVN's extemporised M113 ACAVs ('Armoured Cavalry Assault Vehicles').[63] Helicopters firing TOW anti-tank missiles were also combat proven during the 1972 Easter offensive, and registered a high rate of tank kills. They would have an important part to play in any future war. Another significant development for the infantry in Vietnam was personal body armour, which is now definitely here to stay, extending even to mine-resistant boots. In Vietnam mines were a powerful deterrent to mobility, and the principal cause of US casualties. In the Falklands, ten years later, they were still often undetectable. The mine-resistant boot must hence surely be hailed as a triumph of the offensive spirit, and a turning point in the combat survivability of the foot soldier.

A very different type of question raised by the Vietnam war lay in the whole realm of military ethics and soldierly virtue. During the years of Vietnamisation and withdrawal, from 1968 to 1973, the army had suffered terrible epidemics of demoralisation. Its soldiers were widely denounced not only for the My Lai massacre and other atrocities, but also for drug abuse, Black Power militancy, combat refusal, insubordination and 'fragging' attacks on officers. Abolition of the draft and the introduction of some specific rehabilitation programmes went far towards rectifying these problems,[64] although there have also been more recent complaints that the zero-draft army attracts low-quality recruits who are motivated more by money than by *esprit de corps*. Modern equipment is alleged to be too complex and fragile for its would-be operators in the junior ranks, which suggests that reforms are needed either to simplify the equipment or improve troop quality, or preferably both.

A major difficulty encountered by the army in its attempt to get to grips with these problems has always been its own enormous size, even after the post-Vietnam demobilisation. It is no easy task, even at the best of times, to restructure an organisation of almost a million members. It is still harder to inculcate that elusive quality of 'combat motivation', least of all in times of profound peace. A further American difficulty was that many of the relevant academic studies were written in a strongly sociological style which seemed to hark straight back to McNamara's cost-benefit analysis and business efficiency ethos. An essentially 'humanities' subject was often obscured by inappropriately 'scientific' language, and the impersonality of bureaucracy was only too easily reinforced.

This was of relevance to a second, still more damning wave of post-Vietnam criticism, namely a widespread dissatisfaction with the performance of the officer corps.[65] Not only had officers failed to stave off the general demoralisation towards the end of the war, but they had seemed to connive in a remote administrative system which rewarded peace-time qualities rather than true military efficiency. The individual officer's integrity had been assaulted in Vietnam by the implied pressures of the body count and the efficiency rating system. Too many under-deserved decorations had been awarded to too many 'combat typists' and 'ticket punchers'. Too many officers were too far detached from their men and too 'career-oriented' rather than 'mission-oriented'.

The charge of 'management rather than leadership' refused to go away.

The killing of 231 US marines by a suicide bomber at Beirut airport, 23 October 1983, seemed to suggest that little had really changed. A special commission set up by the Defense Department concluded there had been errors of operational assessment, tactics, intelligence and inter-service co-operation.[66] Nor was the ultimately successful invasion of Grenada in the same year unattended by its own share of mistakes.[67] Seven battalions were needed to overwhelm between 50 and 200 armed enemy, producing nearly 400 casualties to civilians and 80 to friendly troops. Two-thirds of the US casualties were inflicted by friendly forces or by accident – a figure which compares badly with 7% in the battle of the Wilderness, 1864,[68] and surely no more than 25% in Vietnam.[69] Six Blackhawks and twelve other helicopters are thought to have been lost in the Grenada invasion – around 20% of the total number employed – against an enemy armed only with rifles and 0.5 cal machine guns.[70] No surprise was achieved, intelligence and reconnaissance were poor, tactics were still based on 'manoeuvre then fire', and there were grave defects in that hardy perennial, inter-service co-operation. Nevertheless, a staggering total of some 19,600 medals were awarded for the operation, including 8,633 by the army.[71]

There have been widespread calls to reform the professional structure of the US armed forces, and to heighten their military virtues. Indeed, the interest in German military excellence during the two World Wars may be seen as a symptom of this movement,[72] as may American admiration for the British victory in the Falklands in 1982. It is perhaps no accident that the new US infantry helmet has a shape reminiscent of that of the *Wehrmacht*, or that moves have already been made to adopt a 'regimental system' consciously modelled on certain elements of the British structure. American borrowing from European practice is a tradition dating back to von Steuben's drill in the revolutionary war, and it is fascinating to see its revival in modern times.

Any attempt to instil martial values into an advanced twentieth-century society must certainly face some daunting cultural obstacles. At Alamein Montgomery had already been worried about these, and had taken steps to ensure his men were 'worked up into a great state of enthusiasm', while for D Day he wanted them to 'see red'.[73] Much of the basic training for Vietnam was designed to achieve a similar mental attitude, and was largely successful until at least 1968.[74] As we have discovered, however, humanitarian pressures to evacuate the dead and wounded could not be prevented from diverting the course of battles, or producing images for the home front which could only feed anti-war propaganda. Conversely, the battle police did not enforce attendance in firefights in the homicidal manner habitual to the Soviet army even in limited wars like Afghanistan in the 1980s, or the military promenade through Czechoslovakia in 1968. In the American forces combat morale could come only from the inside, but the 'clause of unlimited liability' which any soldier must accept was sometimes quite naturally relegated to a low priority in a 'limited' war.

Perhaps unavoidably, post-war American society was to prove a fertile breeding ground for certain related cults, each of which has subtly changed the way in which military activity is viewed. Natural concern for the missing and prisoners, for example, had already led to such escapades as the Son Tay raid in 1970;[75] but during the following two decades it was to grow into a national obsession of almost mythic proportions. Equally, the combat veteran unable to adjust back to civilian life became an instantly recognised stereotype. Whether he was a psychotic killer to be feared or a suicidal depressive to be pitied, he was generally seen as a direct product of the war and a part of the war's tragedy.[76] The very many combat veterans who became perfectly mature and well-integrated citizens, it seems, have somehow been overlooked in the popular perception.

Contemplation of combat stress itself became something of a cult, as post-1918 revulsion against heavy battle casualties was now bizarrely mirrored in a post-1973 revulsion against battle's mental tensions. After Vietnam everyone thought they knew the war had been terrible, but the actual list of US dead and wounded was relatively short for a war of such duration and importance. Even Korea, where the butcher's bill had been of comparable size, had been compressed into three short years in contrast with more than eight for Vietnam.[77] Apparently 'the horror'[78] of Vietnam had to be sought at a more psychic level, rather than simply in massed physical slaughter. Fear of death; the ironies of fighting for an unseizable, inessential objective; or the guilt of killing non-combatants: these all made more interesting press coverage than either the actual death of American soldiers, or the traditional sombre undertones of straightforward mourning.

Increased interest in the psychological aspects of battle was perhaps inevitable in a society whose children were becoming ever more sophisticated, leisured and sheltered. The sensitive modern man found combat itself a more alien state than had his predecessors, and had to make a more deliberate effort of will to come to terms with it. It was surely no accident that an excessive aura of toughness became attached to élites like the Lurps or Green Berets, who still took battle on the same sort of terms that had at one time been common to whole armies. By 'standing still' they were preserving the ruggedness of past generations, at a time when the modern generation felt itself weakening.

There was another side to the story, however, since it was also commonly believed that a new generation of weapon technology was imposing unprecedented mental pressures upon men in battle. The theory that the men were getting softer was one thing: the apparent hardening of the firepower was quite another. This point emerged strongly from John Keegan's book *The Face of Battle*,[79] which charted the stages by which the 'fear, noise and fatigue' of combat had become successively more stressful through the centuries, up to the point where modern battle would presumably soon become so frightful that it would disappear entirely. His message was one which struck many chords with American readers in the post-Vietnam era.

During the First World War a disease called 'shell shock' had been identified as a medical condition, quite separate from the disciplinary contravention known as 'cowardice'. Fear itself had also been redefined as a perfectly normal emotion for a healthy soldier in combat. Whether for external reasons, or simply because diagnosis had become more fashionable, the incidence of psychiatric casualties thereafter began to climb. In the First World War the British army recognised a level of psychiatric casualties of 4% of its total strength; the German army 0.2%. In 1944–5 it was running at 2.6% of all German troops, 11–12% of British, and as much as 26% of American. In Vietnam there were 2% psychiatric casualties at first, but later rising to a total of 12% serious cases by delayed effect, and a staggering estimate of 80% suffering some form of reaction. In the 1973 October War the Israelis reported between 3.5 and 5% psychiatric casualties, and scarcely less in the Lebanon invasion of 1982.[80] The Israelis prided themselves on particularly good awareness of how the problem might strike, and rapid treatment very near to the front line, which allowed a maximum number of the casualties to be quickly returned to their units.

In the 1980s studies of Soviet tactics were focusing attention on 'continuous operations', which would use night-vision equipment to keep rolling around the clock, imposing severe sleep deprivation on participants. Operations would also probably take place under an NBC threat that would keep each man isolated in his own protective suit or tightly closed vehicle.[81] Where S. L. A. Marshall had wished to break down the isolation of men in battle, therefore, the new warfare of the near future seemed only to be reinforcing it. The general feeling was that this would inevitably make combat harder to bear, although the Airland 2000 concept did at least make a number of constructive high-tech suggestions as to how this might be countered. They ranged all the way from computer games to pass the time, through videotapes of personal counselling by a padre, to chemicals for hygienically disposing of smelly and disturbing human remains.[82]

The massive artillery bombardments inseparable from the Soviet way of war were also identified not merely as a physical danger, but as a very major hazard to psychic welfare. Israeli experience of intense shelling in

1973 was cited in evidence, and the tales of survivors made a sobering impact upon soldiers accustomed to thinking of future combat in terms of dashing manoeuvres and 'death-or-glory' charges. Heavy shelling creates shock waves which numb the senses and cause bleeding from the ears. The danger makes victims' adrenalin run strong; a reaction which in nature is designed to provide the extra energy needed for sharp bursts of life-saving activity. Under shelling, however, there can be no movement, and the victims are forced to lie still under as much cover as they can find, thereby denying the body's demands for violent action to burn out the adrenalin. There is an internal physiological conflict within each man, and he quickly becomes exhausted or breaks under the strain.[83] Added to this are the other horrors of shelling – the dense clouds of dust and fumes; the isolation, once again, from one's comrades; and the appalling prospect of total physical oblivion. A direct hit by a high explosive shell leaves no trace of the person that was its target. There are no remains to mourn, or even to dispose of with hygienic chemicals.

Practical experience of all this had been gained in every major combat from 1915 onwards, and predictions that high explosives would drive men mad had been made as early as the late nineteenth century. Yet western armies seemed to come to it anew in the 1980s, as if it were a specifically futuristic feature of warfare. Just as de Grandmaison and Maud'huy became interested in crowd psychology a generation after Sedan, in fact, so the whole subject of combat stress has gained in its perceived importance a generation after Normandy. Vietnam has spiced that interest with modernity and the immediacy of televisual images, but for the armies preparing to fight *la grande guerre* in Europe, not even Vietnam provides a sufficiently robust model for what may be expected. The likely future shape of large-scale armoured warfare looks just as mysterious and fearsome today as it has ever been at any time in the past.

6

The Recent Military Past and Some Alternative Future Battle Landscapes

Armies have notoriously short memories for the realities of warfare, and despite various attempts to codify and disseminate 'lessons learned',[1] they often become fixated on one particular aspect or procedure and institutionalise it rigidly, while neglecting a broad band of other considerations. Apart from the modern tendency for technology to change relatively rapidly from one war to the next, there are usually changes in the intensity and theatre of the next potential war, which may obscure the lessons from the last. Turnover in personnel is also important, as experienced officers move out of the army, or – no less importantly for doctrine – in and out of particular command appointments and ranks. The British army is perhaps no worse than many others in this respect, but the turbulence it has experienced even since the Falklands war means that it today has remarkably few officers or men performing the same sort of job as they did in battle in 1982 – or even having an influence on how those jobs may be done by others.

A classic case of how a war's lessons may be distorted came in the October War of 1973, when the Israeli army entered some of the fiercest armoured battles seen since 1945, expecting to have a clear field for mobile offensive operations. The war lasted 19 days. In that time, depending on how one counts such things, there were between seven and ten major attacking or counter-attacking moves on the two battlefronts. In the terminology of the Second World War we would say that there had been two 'battles', and at the moment when the war ended the Israelis were in a posture of favourable stalemate on the Golan front, and had won a partial victory on the canal.

The significant feature of all these combats was that, without exception, the defensive side always won *if* it was properly emplaced and organised. It is quite true that three offensive moves did succeed – the

173

Egyptian capture of the Bar Lev Line, the Israeli counter-attack on the Golan, and their canal crossing to encircle the Egyptian Third Army. None of these three attacks, however, met a strong defensive position head-on. To that extent they all had more in common with the early German blitzkriegs of 1939–41 than with the sort of fighting which was seen between 1942 and 1945. Whenever a well organised defensive system *was* encountered, the attack was always destroyed in a pretty conclusive manner.

This result came as a surprise to the Israelis, who had prepared only to fight on the move, as they had in 1967, relying primarily on the firepower and shock of unsupported tanks.[2] Yet the course of the fighting confirmed instead the findings of the last two centuries, in that it demonstrated, yet again, the difficulties which an attacker will encounter against a prepared and 'balanced' defence, while at the same time offering him the hope of triumph in a mobile or 'encounter' battle. At first, however, the results of the October War were interpreted rather differently. Many commentators saw it as conclusive proof that technology – in the shape of the anti-tank or anti-aircraft missile – had come of age. They claimed, somewhat hastily, that new machinery had at last blunted the supposed offensive power of the tank and the aeroplane. Weapons for which their manufacturers claimed a 'one shot kill' capability had apparently managed to overcome air and armoured assaults in the same way that the Maxim gun had overcome the assaults of unarmoured infantrymen in 1915. The power of electronic science, it seemed, had asserted its capacity to dominate ground of tactical importance.

When the Syrians stormed the Golan Heights at the start of the war, for example, they were covered by an anti-aircraft screen which succeeded in shooting down about forty Israeli planes within the first two days. In the Sinai, equally, the Egyptians established a screen of anti-tank missiles around their bridgeheads which defeated all the early armoured counter-attacks against them. On one notable occasion the best part of an Israeli tank battalion was destroyed in a very few minutes and its commander carried as a prisoner to Cairo. It was feats such as these which made a great impression on the world's press.

On closer examination, however, it was found that the new missiles had actually been much less effective than these results would suggest. Far from being 'one shot kill' weapons, they turned out to require multiple firings to destroy their target – and in the case of one famous type of anti-aircraft missile no less than 5,000 rounds were fired to achieve the destruction of only four aircraft. In the tank battle, too, there were numerous stories of tanks coming unscathed from combat with veritable 'cat's cradles' of guidance wires draped around their turrets from missiles which had flown high. Apart from the inherent unreliability of the new missile weapons, it was found that a wide range of counter-measures could be employed against them. As war continued the effectiveness of missiles dropped away dramatically, and more traditional weapons came to the fore. Against aircraft the greatest successes were scored by multiple heavy machine-guns. Against tanks the bazooka and the tank gun itself did much more damage than all the missiles put together.[3]

Just as in the Second World War, tank commanders came to realise that they could not move forward without the close support of infantry, artillery and other arms. Before 1973 the Israeli tank corps had confidently relegated all other arms to a subordinate role; but now they were forced to reconsider this doctrine overnight. As at Alamein, it was found that the open desert terrain could be held by a relatively few defenders against all but the most methodical and well co-ordinated attacks. Far from representing a technological break-through, the October War soon began to look like a re-affirmation of past lessons.

The Israeli army next went into action in 1982 in 'Operation Peace for Galilee', fought in a Lebanon where there had not been, and would not be, peace. An extensive attempt was made to apply the lessons of 1973, especially insofar as there was now much more inter-arm co-operation and a greatly expanded artillery. There was close infantry support for the tanks, and a willingness to use 'prophylactic firepower' against potential sources of anti-tank missiles or rockets. Technology was certainly harnessed with gusto, including sophisticated electronic warfare in the air, realtime RPV surveillance in the ground battle, much increased use of anti-tank helicopters, and a new Merkava tank with reactive armour and innovative systems to damp down internal fires. The Israelis lost only one aircraft, two helicopters and some 50 tanks, but sliced through the Syrian and PLO forces in short order, destroying 86 aircraft, all the air defences, and some 350 tanks including nine of the widely-feared Soviet T-72s.[4]

On the other hand, the Israelis still found they were short of infantry, and were unprepared for some of the conditions they encountered. Beirut offered the prospect of fighting for a city, for which they had as little previous experience as had those other exponents of rapid armoured manoeuvres – the Germans when they first approached Stalingrad in 1942. In 1982 the Israelis chose not to repeat Paulus' mistake, although by merely sitting back and using artillery they achieved little beyond attracting international odium. Nor were they fitted for the counter-insurgency warfare which followed. Prophylactic firepower is not an appropriate weapon for winning hearts and minds in built-up areas, and conscript soldiers trained for *la grande guerre* do not convert easily to the subtleties of peace-keeping.[5]

If the stresses of fighting in populated areas were new to the Israelis in 1982, there had also been some severe psychological tensions in 1973.

The wide dispersion of the defending force and the very long engagement ranges (in daylight at least) threw individuals back, once again, on their own resources. Sometimes under ferociously heavy shelling and often outnumbered by up to ten to one at the point of contact, the soldier felt the strains of battle weighing upon him very heavily indeed. We therefore find that after this war, as after so many others, the tacticians were at pains to stress the importance of finding soldiers of high quality. Realistic training to fit a few picked men for battle was seen as preferable to a system which trained more soldiers less well.[6] This idea can scarcely be called revolutionary.

Nor have different lessons been learned in other recent conflicts. The most celebrated case is the Falklands war of 1982, when the British landing force, consisting of infantry on foot with only relatively light artillery support, defeated twice its own number of dug-in defenders. It won a classic victory for training, unit cohesion and military professionalism, over poor Argentinian leadership and badly-motivated conscript soldiers. Some technological innovations were certainly highlighted by the campaign, such as the significant success of Harrier jump jets firing Sidewinder missiles in air-to-air combat, but generally the striking tactical achievements lay more in the realm of small-unit resilience, responsiveness and determination. Machine gun positions dominating bald hillsides had not cheaply been overrun in Flanders or Normandy during the two world wars, when casualties often reached between 20% and 70% in battalion attacks which ultimately failed. At Goose Green in 1982, however, the 2nd battalion of the Parachute Regiment lost only 52 casualties (i.e. 11% of its strength) in an utterly successful 36-hour 'battle of annihilation' against twice its own numbers backed up in depth positions.[7]

The Falklands may perhaps be dismissed as too much of an 'air and naval'[8] or 'limited' war, with opponents too unequally matched to make a 'fair' or 'representative' fight. However that may be – and the calculus soon becomes bizarre – it scarcely featured a major clash of armour and helicopters. The Iran-Iraq Gulf War, in contrast, did see this on a large scale. So evenly matched were the combatants, furthermore, that for almost nine years neither one of them could push its offensives home, and a peace of exhaustion finally had to be declared. For the Iraquis there was the frustration of finding that their initial lightning all-arms thrust could be stopped in its tracks by determined militia deployed in depth. For the Iranians there was a repeated failure to launch a successful counter-attack, despite great advantages of numbers. By the end of the war they are estimated to have lost something like a million dead, mostly concentrated on a battle front little more than a hundred miles long.

The war was one of attrition and fixed fortification, where the use of carth-moving equipment took on an enormous importance in the flat and featureless landscape. It was not entirely a static war, however, since relatively small numbers of well-trained and well-equipped Iraqui defenders were often able to use skilful manoeuvres to contain, counter-attack and destroy large Iranian assaults. Used in mass, the helicopter came into its own as a mobile 'fire brigade' force that could rush to a threatened point and blunt the onslaught. Admittedly, the terrain was exceptionally inhospitable for the attacker, whose mix of weapons and training also usually left much to be desired: nevertheless, professionalism in husbanding scarce resources was a central feature in Iraqui survival.[9]

The Gulf War confirmed the lessons of other recent conflicts insofar as it showed the enormous logistic strain imposed by modern operations, no less than the difficulties of penetrating a prepared belt of defences using troops of less than excellent quality. In Afghanistan the hardy *mujahadin* encountered similar frustrations in their own assaults, although their mountain strongholds proved inaccessible, in turn, to the road-bound Russians – apart from specialist units of commandos with helicopters. Just as the Gulf War reproduced the attrition of the trenchlock on the Western Front, so the Afghan war reproduced the attrition of Vietnam. Ancient military values re-emerged as important as ever, and modern technology could never provide more than a part of the answer.

The non-appearance of a true revolution in tactics should not by now surprise us, since we have already found several other supposed tactical 'revolutions' which turned out to be nothing of the kind. We have also seen many cases of bright new technology failing to deliver what it promised, and the decision reverting to the quality of individual troops. To that extent the October, Gulf and Lebanese Campaigns fit easily into the pattern established in the past. If we consider the ranges of engagement in open terrain, on the other hand, we do find one significantly new feature which seemed to be emerging. Put simply, we find that in these wars there was astonishingly little close combat.

The prevention of close combat, however, has always been one of the primary functions of weaponry. By killing the enemy at a distance and in numbers one is able to put off the sickening moment of personal confrontation face-to-face. One can limit one's personal exposure to danger and decrease the effect of chance upon the outcome. Instead of plunging into a roughly even contest of man against man, the warrior with the long range weapon can hover tentatively around the perimeter of the fighting. He keeps open the option of flight, or at least that of personal protection.

To this extent improvements in weaponry tend to limit the intensity of the fighting rather than to increase it, and it is a remarkable paradox that the practical lethality of weapons has actually fallen away sharply over the past two centuries. In Napoleonic battles it was calculated that about three to five hundred shots had to be fired for each hit on an enemy soldier, whereas in the Manchurian war the number was nearer 20,000 shots, and in Vietnam around 600,000. In terms of Israeli history we find that in the highly mechanised fighting of 1973 and 1982 they suffered many fewer casualties than they had in the purely infantry battles of 1948. On the latter occasion the two sides did not have the heavy weapons which might have kept them safely apart, so they were forced to enter the dangerous and unpredictable business of close quarter combat.

When measured by the rate of casualties, battle in general has become less and less intense as weapons improve. Napoleonic soldiers might manoeuvre for weeks or months before fighting for only a few hours; but when they did fight it was not abnormal for armies to suffer 20–30% casualties, and regiments at the spearhead to suffer 70–80%. The combined casualties of the two sides engaged at Waterloo were some 68,000 and at Borodino 75,000, each in a space of twelve hours. In the four-day battle of Leipzig there were 127,000 casualties, or 32,000 per day. As weapons improved, however, the rate of losses started to decline. In 'America's bloodiest day' at Antietam, 1862, the combined total loss was 26,000 in twelve hours; at Gettysburg, 1863, it was 43,000 in three days, or 14,000 per day; at Mars-la-Tour/Gravelotte in 1870 it was something like 66,000 in a three-day episode, or 22,000 per day.[10]

By the time we come to twentieth-century battles we find that, apart from a very few 'near-Napoleonic' days, such as the 60,000 lost on 1 July 1916, the graph continues to fall. A casualty total of 750,000 in ten months at Verdun averages less than 2,500 per day; half a million in 100 days of Passchendaele makes just 5,000 per day, and about a million lost in 140 days on the Somme averages around 7,000 per day. These rates are comparable to some 40,000 in twelve days at Second Alamein (including POW) making 3,300 per day, plus a daily average of 83 tanks knocked out. In Normandy the figures were 637,000 casualties during 80 days, making 8,000 per day of which perhaps 1,100 might be killed, 4,300 might be wounded and 2,600 POW. Tank losses averaged about 30 per day. The major exception to this general level was in Russia during the Second World War, where it seems that everything was on a greater scale. A total of 30,000 men were lost per day during the six-month drive to the outskirts of Moscow in 1941, and the Russians alone lost some 13,000 per day in the final battle of Berlin.[11]

We have already seen how relatively light were the American losses in Vietnam, which work out at an average of only sixteen men killed (but an estimated $50 million spent) each day between deployment in 1965 and final ·withdrawal in 1973 – although the total combined casualties (including wounded and missing for both sides) may have been as high as 1,000 per day. Israeli losses averaged up to 158 killed per day in 1973, and perhaps 68 per day in 1982; but once again with much higher combined overall totals, at something like 5,000 and 1,700 per day, respectively. The combined tank losses of the two sides were higher still, at around 142 and 60 per day, respectively. British casualties in the Falklands totalled 255 dead and 777 wounded in 42 days' operations, as against about 10,000 Argentinians (mostly POW). This makes an average of 6 British dead per day, or 263 total casualties from both sides[12] – not even a third of the daily rates in Vietnam.

The above figures, of course, take no account of the numbers initially deployed, or the precise definition of just what constitutes a 'battle'. Even if we discount the post-1945 examples as being too 'limited' to bear comparison with *les grandes guerres* of earlier times, however, the general trend of casualties per day of fighting is clearly downwards. Soldiers facing improved weapons take more care to hide or protect themselves, and engagements become more fragmentary. Even if much greater total weights of ammunition are expended, the really damaging fire missions become more intermittent and infrequent.

Ardant du Picq was right to say that:

In modern battle, which is delivered with combatants so far apart, man has come to have a horror of man. He comes to hand-to-hand fighting only to defend his body or if forced to it by some fortuitous encounter.[13]

Close combat, in other words, tends to take place by accident, when the technology fails which might otherwise have prevented it. It is not generally desired by either party, since it represents a removal of psychological restrictions upon aggressive behaviour. It comes closer to the essence of total war than can any amount of fighting at long range.

When and if they do find themselves face-to-face with an enemy, soldiers may still attempt to limit the intensity of the combat by recourse to a form of tacit mutual understanding:

During the Crimean War, we are told, on a day of heavy fighting, two detachments of soldiers, A and B, coming around one of the mounds of earth that covered the country and meeting unexpectedly face to face, at ten paces, stopped thunderstruck. Then, forgetting their rifles, threw stones and withdrew. Neither of the two groups had a decided leader to lead it to the front, and neither of the two dared to shoot first for fear that the other would at the same time bring his own arm to his shoulder.[14]

For a more modern example of the same thing, compare the case of an American officer, Captain Willis, leading his company along a Vietnamese streambed one night in 1966:

Two bullets whistled over Willis' shoulder, fired from not more than ten feet away. He did not return the fire. . . . One [enemy] soldier had awakened and, hearing the sounds of movement, had gotten off two quick shots, probably not even looking.

179

For the next few seconds the numbers of the forms passing him in the stream must
have awed him somewhat.

Willis came abreast of him, his M-16 pointed at the man's chest. They stood not five
feet apart. The soldier's AK47 was pointed straight at Willis.

The captain vigorously shook his head.

The NVA soldier shook his head just as vigorously.

It was a truce, cease-fire, gentleman's agreement or a deal. Call it what you will.
At least the understanding of the compact was mutual and complete.

'If you don't try to fire again and alarm the camp, nobody here will try to kill you.'

The soldier sank back into darkness and Willis stumbled on. . . .[15]

In view of the great reluctance of soldiers to mix it hand to hand, it has
long been recognised that the side which goes out and actively seeks a
confrontation will enjoy a great psychological advantage. Provided that
the enemy can be convinced of both your intention and your ability to
reach him, he will in all probability run away and leave you the victory.
This was the secret behind Wellington's successes in the Peninsula, and
it continued to be decisive on many occasions throughout the nineteenth
century. With improvements in firepower, of course, it became
gradually more difficult, both physically and morally, to reach an
enemy's line. Looser formations and heavier fire preparations had to be
used in the bayonet charge, while those responsible for training were
ever on the lookout for better ways to stiffen the initiative and resolve of
their men.

In the twentieth century a similar process has steadily continued.
Formations have become looser still, and the battlefield has become even
emptier. As a result personal initiative has loomed ever larger as an
essential military virtue which it is the task of training to develop.
Whereas in the past this had been done by reference to pseudo-scientific
principles, in the Second World War and Vietnam some armies even
brought in real scientists – especially psychologists – to help them. The
theories of 'morale' which had been so despised in their late nineteenth
century guise now reappeared in bright new 'rational' clothing. The idea
of close combat also continued, especially after it had been given a new
lease of life by the storm troopers of 1917–18. The quick counter-charge
remained as much a part of defensive technique as it had for the Duke of
Wellington; while if attacking techniques had become more complex
they still demanded soldiers who would 'go'.

In numerous wars since 1945 by contrast, we suddenly find that for
different reasons the idea of close combat has apparently been lost. In
Vietnam the Americans' policy of 'Manoeuvre and Fire' seemed to
prohibit it, and in the Middle East the terrain and the weaponry made it
generally possible to break up attacks at long range – often a kilometre or
more. In these circumstances we might well be permitted to wonder
whether technology has at last succeeded in rendering no-man's-land

uncrossable, and whether the bayonet has at last become redundant. It is certainly noticeable that small arms designers today make markedly less provision for the bayonet than at any time since the War of the Spanish Succession.

As we have already seen, however, the history of the bayonet has been the history of its premature obituaries. Even in Napoleonic times, when it was in its glorious heyday, it had to suffer many predictions that it was already obsolete.[16] Throughout the nineteenth century and into the twentieth the same cry has often been repeated[17] . . . and yet the close combat and even the bayonet charge itself have continued undiminished until apparently, the wars since 1965.

Admittedly the bayonet today has been largely replaced by other close-quarter weapons such as the hand grenade or the SMG; but the real question, perhaps, is concerned more with the 'spirit' of the bayonet than with the weapon itself. Are we to conclude that troops no longer need to make the final risky step across no-man's-land, if they are to win the victory? Are we to accept that long-range technology has finally become decisive? On past performance it would seem unlikely that this change has really occurred. Perhaps there were special features in Vietnam and the Middle East which made it seem that a real change had taken place when it had not. The North Vietnamese Army, after all, continued to believe in the attack *à l'outrance* and in the event they won their war. In the Middle East the terrain was exceptional, as deserts have always been for tacticians. Even Napoleon's *grognards* seem to have set aside their 'battles of Egypt' as something extraordinary: for them these were a completely different type of action from other battles nearer home.

For all this we have still, at the backs of our minds, a dark suspicion that in fact the close-quarter combat really is a thing of the past. We have a deep-seated feeling that the cold-blooded technician now controls the battlefield, and that the hot blooded warrior is an anachronism. Tactical analysts such as Mao Tse Tung, who put 'men above machines', are easily dismissed as 'military spiritualists' or 'military Luddites'. For all their famous victories, they are deemed to have been superseded by the march of science.

This process may perhaps be seen at its most intense in the preparations for a possible future war in Germany. There the uncertainty of the unknown itself encourages commentators to look more at tangible things – the hardware and solidly scientific weapon power – rather than more traditional but less tangible human values. When they describe recent real wars in the Falklands or the Middle East, analysts are often able to balance bald, abstract science with a variety of telling human insights and anecdotes, since they are discussing events that are already past and gone. When it comes to describing the shape of the next big state-of-the-art war in Europe, however, there are too many dollars and reputations hanging on the result for similar rules of evidence to apply.

The debate about the future of warfare is inevitably far more political than the debate about the past. The issues seem far too important to be blurred by such apparently vague factors as unsubstantiated subjective feelings about how particular men may behave. For example, inter-allied discussions of strategy may not be conducted in terms of comparative regimental cohesion or personal devotion, since these are 'merely' indefinable, unquantifiable virtues which not even the recent interest in combat stress has made more concrete. It would cut no ice at a NATO planning conference to suggest that the 93rd Sutherland Highlanders might possess stronger military qualities than the US 385th Air Mobile Cavalry Regiment, because all that can be discussed is what can be counted, and all that can be counted is the number of helicopters, guns and rockets they can each bring along with them to the next battle. Such hardware and logistical considerations do of course have an important role to play, and no one would wish to deny it – the Highlanders would certainly like to have more firepower than they actually have! However, the very nature of debate about future warfare often tends to over-emphasise those aspects at the expense of more important but less quantifiable things.

The bewildering technical complexity of modern military science is in any case itself a source of uncertainty enough. Already in recent wars we have seen countless new types of assault helicopter, guided missile, smart artillery sub-munition and scatterable minefield. New laser designators have been used to find targets for both artillery and air attacks, while new laminated and reactive armour has been seen and even defeated. Numerous advances have also been registered in electronic warfare, remotely piloted vehicles, secure communications, realtime battlefield surveillance, target acquisition and reconnaissance ('STARS'). The feeling is strong that the micro-chip has really revolutionised warfare as much as did the internal combustion engine in its day.[18] There are many still more significant developments waiting just around the corner, not least of which is a family of directed energy weapons that may make battlefield death rays a reality by the year 2020. In these circumstances we may perhaps be forgiven if we find it difficult to attribute a correct value to each of the new systems, let alone form a useful view of how the battlefield as a whole may appear.

One theme which seems to run through most literature on future tactics is that the 'emptiness' of the past battlefield is declining, in that combat units are becoming ever more visible to the enemy. There are many more ways to see men and weapons than there used to be, making traditional camouflage and concealment methods ineffective. Whereas a piece of netting entwined with green scrim was enough to hide a gun position or a vehicle in the past, in future there will have to be additional reflectors to defeat various different types of surveillance, and probably also a set of alternative radiation sources to act as decoys. Still more significant, the cover of darkness or woods will no longer hide vehicles on the move. Innovations will have to be made in concealment and ECM technology if we are to emulate Balck's surprise night marches with 11th Panzer Division on the Chir in 1942.

A highly probable solution to decreasing concealment, however, is to restore the 'empty' battlefield by increasing dispersion. If large formations can easily be noticed and monitored, the use of smaller ones may well become unavoidable, and all the more so since small units today can pack a much bigger punch than they could in the past. If a tank before the 1970s needed to fire four or five shots to hit an enemy tank, the figure is now reduced to just one or two shots.[19] This presumably implies that a squadron today can almost do the job of a battalion yesterday. An all-arms force with as many sub-units, technical support and command elements as a brigade, but perhaps only half its present 'teeth' strength, may well become the basic tactical unit of the future. This would bring many advantages in terms of economy of force, and would finally lay to rest the 1941–4 debate about brigade groups or divisions. Neither of those formations is probably now small enough to survive easily or inconspicuously on the battlefield.

The logic of ever-longer weapon ranges also continues to drive armies towards increased dispersion. Whereas most direct fire anti-tank weapons in 1945 were effective only below 2,000 metres, the average today is 3,000 metres or more, with as much as 10km being possible in some cases. Sufficiently extensive fields of vision, however, are unlikely to be found except in desert or steppe conditions, and it may not be coincidental that there has recently been a great emphasis on the use of indirect fire weapons against armour. Whereas field artillery in the two world wars usually had an effective range of ten kilometres or less, however, this zone can now be fully covered by mortars – including automatically loaded and 'smart' anti-tank varieties – while most gun and rocket systems in use today can reach out to 20km or more. Ranges around 30km are attainable not merely with the long Soviet 130mm or American 175mm guns, and with both the BM-27 and MLRS multiple rocket systems, but also with many modern self-propelled 152 or 155mm guns if they fire base-bleed or rocket-assisted projectiles. Still deeper into the enemy's territory, interdiction by manned aircraft has now been joined and complemented by a variety of surface-to-surface rockets carrying conventional payloads.

At least in theory, a standard field artillery battery should soon be able to cover a frontage that in 1945 would have required three or four batteries, and with a depth of reach that was attainable only by aircraft or certain very specialised guns. A multiplicity of novel ammunitions will also allow armoured vehicles, radars and command posts to be selectively and appropriately engaged. If greatly enhanced target acquisition, accuracy and responsiveness are further weighed in the balance, it is not too fanciful to talk of artillery soon multiplying its general effectiveness for mobile warfare by as much as ten times since 1945. This must admittedly be set against an equal and opposite increase in the efficacy of counter-battery fire, which will force firing units to keep hidden or keep moving; but a very major increase in artillery power will still surely have become general within the next two decades. In that time we will also doubtless see a shift away from existing solid propellants to the lighter, cheaper and handier liquid propellants.[20]

The speed at which forces may be inserted into the battlefield has also increased since the world wars, not only as a result of greatly improved communications, but also through better and faster cross-country vehicles, especially helicopters and hard-hitting light armour. This means that a defender may be able to thin out the forces manning his front line still further, relying on rapid reinforcement, if necessary, only at the last minute. The capacity to fill gaps by scattering minefields rapidly from a distance – by artillery, helicopters or fixed-wing aircraft – is also highly significant in this context. We may thus perhaps be facing a situation similar to that of the Western Front in the First World War, where mobility was easy behind the line, but very difficult in proximity to the enemy.

Can we really decide just what sort of battle landscape would apply to Germany, as a result of all this increased dispersion? We may certainly expect the ground to be less obviously cluttered with military equipments, and to show fewer scars than was the case in the past. There will not only be fewer tanks and trenches per square mile, but fewer and smaller shells aimed against them, in shorter but more precise bursts of fire. Tactical measures and deployments, however, will probably be necessary scores of miles behind the FEBA, making for a much deeper battlefield than in the past, with a less clear-cut distinction between 'military' and 'civilian' zones even than was seen in 1945. Civilian machinery will also be exploited to a greater degree; not just for support tasks like transport and the telephone net, but in the firing line itself. Earth-moving plant will be requisitioned to dig troops into position, and motor vehicles to create heat sources and decoys. There may well be an attempt to maintain some of the patterns and intelligence signatures of normal civilian life, so that the armies may blend in more easily. An armoured brigade is far harder to conceal in areas where central heating, cooking and television have all been turned off, and local traffic stopped, than where the population is still going about some version of its everyday business.

Many projections of a conventional war for Germany have anticipated massive 'collateral damage' and the wholesale destruction of cities. This may indeed be the case at specific points of main effort, where HE shelling is especially concentrated. There would also be very widespread demolitions to block roads and defiles. Nevertheless, neither side would wish to fight large battles in towns if they could be avoided, and a speedy conclusion to operations would certainly be preferred to stand-up attritional fighting. The more discriminating and economical nature of much modern ammunition may even mean that a short war lets the landscape off relatively lightly – apart perhaps from a persisting carpet of millions of unexploded minelets.

A major unanswered question – if not the most important question of all – concerns the speed at which a battle line might be expected to move backwards or forwards. Both sides would hope to 'bite fast and deep', in de Grandmaison's phrase, keeping up the momentum of their spearheads by making overwhelming concentrations at key points. However, the logic of increasingly dispersed tactics would suggest that such concentrations might today be prohibitively dangerous. The statistical norms for breakthrough ratios, which might have worked with effortless ease in the Great Patriotic War, may now lead to disasters still greater than the attempted break-outs at 'Goodwood' in 1944 or on the Golan in 1973. Unless deep force reductions have been negotiated and accepted before a war for Germany starts, a defender may always be able to find the squadron, the company – or the force equivalent in heli-borne reserves or scatterable mines – that he needs to contest every avenue of approach. The range and flexibility of firepower, combined with realtime surveillance by HQs, should then allow him to reinforce each of the threatened points very rapidly, as the Iraquis were able to do in their repeated Gulf War defensives. The gaps could then be plugged and the battle front stabilised into a frontal deadlock.

The logic of the dispersed battlefield, however, also implies that devastating infiltrations by small units may still be eminently possible for an attacker. Given the right type of laser designator and rear-link transmitter, it may need only one observer lurking fifty kilometres behind enemy lines to bring down instant pinpoint fire against a key HQ or ammunition dump. Two or three raiding helicopters or armoured cars may be able to slip through narrow gaps in the front, to take out the supply echelon for a brigade. A fast-moving force of a dozen or so tanks may be able to manoeuvre around an enemy's flank to surprise and destroy an armoured battalion within a few moments. A highly fluid and 'intermingled battle' is thus a distinct possibility in the close country of Germany,[21] and we may find ourselves looking more to the Vietnam model of forest infiltration and mobility than to the Gulf War model of locked fronts in open terrain.

This type of battle may well be enhanced if the new surveillance and firepower technology fails or can be countered, as was already often happening in Vietnam. To meet the new electronic weapons, the number and sophistication of new electronic counter-measures is today certainly prodigious, albeit often shrouded in secrecy and uncertainty. They include such things as jammers fired in shells against enemy HQs to obliterate transmissions locally, and counter-laser lasers to decoy incoming beams or blind their operators. Smart munitions may also be vulnerable to quick-reacting high-volume point defence systems, designed to shoot down bombs and missiles before they can effectively engage. Smokes, chaffs, flares, anti-radiation missiles and simple but stupendous jamming energies all have a part to play, and more than a few commentators have wondered whether command, contrɔl and target acquisition may not all break down completely under the pressures of real battle. Even without having to contend with ECMs, indeed, many electronic weapons carry built-in weaknesses simply because of their own inherent complexity. A system which uses eighteen different technologies is surely at least six times more likely to fail than one which uses just three.

A dispersed but fluid and 'intermingled' battle might be embarrassing for NATO's somewhat linear concept of 'forward defence', and it would entrust the ultimate decision to the accidents of personal inspiration and training at relatively low levels of command. Nevertheless, it is a type of contest which the military establishments on both sides believe they can win, and which even NATO would in many ways prefer to a more predictable trenchlocked battle of attrition. Everyone agrees that the best war is one that can be 'over by Christmas', especially since no one has stockpiled ammunition to fight for more than a few weeks. Yet it remains as true today as it was on the Marne in 1914, or on the Gulf in 1980, that neither side can win quickly, even in the most mobile of wars, if both sides fight equally well and survive the initial phases. Conversely, a rapid decision can be reached even in the most static of trench deadlocks, provided one side has an overwhelming advantage over the other. It is therefore the duty of analysts to discover just where such an advantage may lie, and in this process a judicious understanding of the military past can be no less rewarding than a mastery of the technology and hardware.

Notes

Chapter 1

1 There are only a few alternatives to written evidence for the history of tactics. One of them is battlefield archaeology: see e.g. Scott, D. D. and Fox Jr, R. A., *Archaeological Insights into the Custer Battle* (University of Oklahoma, 1987), but also the many straightforward guidebooks for battlefield tourism. Another approach is oral history: for example, the systematic interviews with world war survivors that have been tape recorded by the Imperial War Museum, London; but see also the many diverse volumes of interviews and battle impressions that have been published in recent years in Britain and USA.

2 Lasswell, H., *Propaganda Technique and the World War* (New York, 1938), p. 90.

3 Convenient summaries of the evolving tactical debate may be found in e.g. Paret, P, ed., *Makers of Modern Strategy* (Princeton, 1986); or Wintringham, T., and Blashford-Snell, J. N., *Weapons and Tactics* (London, 1973).

4 See especially Appendix I, *The Art of Tactical Snippetting* in my *Rally Once Again* (Crowood, 1987, and – as *Battle Tactics of the Civil War* – Yale University, 1989), p. 193 ff.

5 Colin, J., *L'Infanterie au XVIIIe. Siècle; la Tactique* (Paris, 1907). See also Léonard, E. G., *L'Armée et ses Problèmes au XVIIIe. Siècle* (Paris, 1958).

6 Hamilton, I., *A Staff Officer's Scrap Book during the Russo-Japanese War* (2 vols., London, 1905), vol. 2, p. 194.

7 Siborne, H. T., *Waterloo Letters* (London, 1891), p. 383.

8 See the notes to Chapter 5, below. A controversial and stimulating thesis on the nature of battle reporting is Keegan, J., *The Face of Battle* (London, 1976).

Chapter 2

1 Jomini, H., *Précis de l'Art de la Guerre* (new edn., 2 vols., Paris 1855. The first complete edition appeared in 1838.), vol. 2, p. 231; and *Précis Politique et Militaire de la Campagne de 1815* (Paris, 1839), p. 205.

2 Oman, C. W., *A History of the Peninsular War* (7 vols., Oxford, 1902–1930), and the articles referred to below in footnote 34. Colin, J., *La Tactique et la Discipline dans les Armées de la Révolution* (Paris, 1902), and *Les Transformations de la Guerre* (Paris, 1911).

3 Quimby, R. S., *The Background of Napoleonic Warfare* (New York, 1957) is an even more extreme translation of Colin's work; while Oman's ideas have been given a late twentieth century flavour by two weaponry experts – Jac Weller in his histories of Wellington's campaigns, and Major General B. P. Hughes in his *Firepower* (London, 1974).

4   Weller, J., *Wellington in the Peninsula, 1808–1814* (London, 1962), p. 47.
5   Gurwood, ed., *The Dispatches of the Field Marshal the Duke of Wellington* (London, 1835), vol. 4, p. 95.
6   Curling, H., ed., *Recollections of Rifleman Harris* (London, 1929), p. 50.
7   Wyld, J., *Memoir Annexed to an Atlas* (London, 1841), p. 4.
8   Landmann, Colonel, *Recollections of my Military Life* (2 vols, London, 1854), vol. 1, pp. 212–14.
9   Fyler, Colonel, *History of the 50th (or the Queen's Own) Regiment* (London, 1895), pp. 105–6.
10 Anon., *Recollections of the Peninsula* (5th edn., London, 1827), p. 151.
11 *USJ*, September 1844, p. 92.
12 *USJ*, June 1834, p. 183.
13 Siborne, H.T., *Waterloo Letters* (London, 1891), p. 383.
14 *USJ*, October 1829, p. 417.
15 *USJ*, June 1834, p. 183. The British cheer was already thought to be more effective than volley fire at the battle of Dettingen, 1743: Orr, M. J., *Dettingen 1743* (London, 1972), pp. 65–6.
16 See Nafziger, G. *The Thin Red Line: A Tactical Innovation or a Circumstantial Necessity?* in *EEL* no. 62, March 1982, pp. 4 ff. Note that the restricted frontages of many British battles lead one to speculate that in practice the troops were often packed more closely than two deep.
17 *USJ*, February 1830, p. 207. Michael Glover's inspection of the ground at Albuera, however, has convinced him that no such gulley existed (private communication to the author).
18 Two of these battalions were probably composite units made up of two weak battalions each, hence a total of 'seven battalions' is sometimes given. See also Shopfer, P. A. & Lochet, J. A., *The Attack and the Formations of the Middle Guard at Waterloo* in *EEL* no. 75, October 1983, pp. 38–9, summarising a lengthy debate. Compare also J. E. Koontz' remarkable analysis, *Note on D'Erlon's First Attack at Waterloo* in *EEL* no. 78, March 1984, pp. 47–55, and no. 79, April 1984, pp. 19–44.
19 *Waterloo Letters*, op. cit., p. 248.
20 Ibid., p. 277.
21 *USJ*, March 1845, p. 403.
22 *USJ*, April 1845, p. 471.
23 *USJ*, June 1841, pp. 179–180.
24 Quoted in Gover, M., *Wellington's Army in the Peninsula, 1808–1814* (Newton Abbot, 1977), p. 169.
25 *USJ*, June 1841, p. 181.
26 Ibid., p. 183.
27 *USJ*, July 1834, p. 321.
28 *USJ*, January 1848, p. 41.
29 A. F. Becke, quoted in Quimby, op. cit., p. 342.
30 *USJ*, December 1833, p. 438.
31 Ardant du Picq, C.J.J.J., translated by Greely, J. N., and Cotton, R. C., *Battle Studies* (New York, 1921), p. 151.
32 Fortescue, J. W., ed., *Memoirs of Sergeant Bourgogne, 1812–13* (London, 1926), p. 8.
33 Saint-Pierre, L. de, ed., *Les Cahiers du General Brun de Villeret, Pair de France, 1773–1845* (Paris, 1953), p. 146.
34 Oman, C. W., 'The Battle of Maida' in *Studies in the Napoleonic Wars* (London, 1929. The article on Maida first appeared in 1908); and 'Line and Column in the Peninsular War' in *Proceedings of the British Academy*, London, vol. 4, 1910, later reprinted in *Wellington's Army, 1809–1814* (London, 1913). Maida and its histori-

ographical implications are discussed in Arnold, J., *Column and Line in the Napoleonic Wars. A Reappraisal* in *Journal of the Society for Army Historical Research* vol. LX, 1982, pp. 196–208.

35 Anon. ('An artillery officer commanding a French battery at Maida'), 'Combat de Maida, Rectification d'une Erreur de Walter Scott' in *Spectateur Militaire*, Paris, No. 23.

36 Colin's preface to *La Tactique et la Discipline*, op. cit. Modern research tends to confirm these findings, at least in the period before 1808.

37 *Archives Historiques de la Guerre (AHG)* at Vincennes; Carton C 7/6, Junot's MSS report 31st August 1808, p. 8. and Thiébault's MSS report 15th September 1808, p. 16. Carton C 8/147, Soult's MSS report 18th May 1811, p. 198.

38 *Précis de l'Art de la Guerre*, op. cit., vol. 2, p. 231.

39 Martin, *Souvenirs d'un Ex-Officier, 1812–15* (Paris, 1867), p. 160.

40 Chambray, G. de, *De l'Infanterie* (Montpellier, 1824), pp. 16–17.

41 Quoted in Maxwell, H., *The Life of Wellington* (4th edn., 2 vols., London, 1900), vol. 2, p. 320, footnote.

42 Colonel Hanger's famous remark is quoted in *e.g.* Gates, D., *The British Light Infantry Arm, c. 1790–1815* (Batsford, London 1987), p. 139.

43 E.g. Duhesme, C., *Essai Historique sur l'Infanterie Légère* (first published 1814, 3rd edn, Paris 1864), pp. 145–9, 302. Duhesme's work is often overlooked, but was a very important inspiration for many French tactical writers in the nineteenth century.

44 Weller, J. *Shooting Confederate Infantry Arms, Part One* in *The American Rifleman*, April 1954, pp. 43–4.

45 Busk, H. *The Rifle* (first published in the mid-1850s. 4th edn., London, 1859, reprinted Richmond, England, 1971), p. 18. The well-known Vittoria statistic seems to have appeared first in Henegan's 1846 memoirs vol. I, pp. 344–6, cited in e.g. Gates, op. cit., p. 140.

46 Nafziger, G., *French Infantry Drill, Organization and Training* in *EEL* 39, October 1979, p. 17. Guibert had claimed an even longer maximum, at 600 *toises*, or 1170 metres, in *Essai Générale de Tactique* in Menard, ed., *Guibert, Écrits Militaires 1772–1790* (Paris 1977) p. 124. The comparable figure for the British Brown Bess was apparently 700 yards: Haythornthwaite, P., *Weapons and Equipment of the Napoleonic Wars* (Blandford, Dorset, 1979), p. 19.

47 German target scores quoted in Lauerma p. 32, and Scharnhorst *Uber die Wirkung des Feuergewehrs* (Berlin 1813, new edn., Osnabruck 1973), pp. 93–7. Compare comparable British and French results discussed in Haythornthwaite, P., op. cit., pp. 19–20.

48 French infantry inspection reports for 1831–45, covering some 368 separate shoots, in AHG cartons Xb 626–726. Busk, op. cit., gives an astounding example of 11% hits being achieved with French muskets at 600 metres!

49 Holmes, R., *Firing Line* (Cape, London 1985) p. 168, based on Fortescue. The small number of troops in the sample, however, make this a result for 'one tactical action' rather than for 'a whole battle'.

50 See my notes in *EEL* no. 101, July–August 1987, pp. 4–7.

51 See my comments on Bressonnet's *Études Tactiques sur la Campagne de 1806* (Chapelot, Paris 1909), in *EEL* no. 82, September 1984, pp. 16–22, showing how French skirmishing was already widespread as a 'decisive' combat technique.

52 Gates, op. cit., pp. 138–48, for the movement towards aimed fire, from the 1790s onwards.

53 In the French range tests of 1831–45 already cited, 'light' infantry regiments were not noticeably more accurate than line. The 6th Light in 1842 scoring a record low of 0.6% hits, as opposed to the staggering claim from 37th Line in 1844 that it had achieved 70%! Overall only 46 of the results (12.4%) showed a score higher than 20% hits; but only 30 of them (8.1%) registered 6% or fewer. Note also that in four other comparable cases

the *chasseurs à pied*, using rifles, succeeded in scoring an average of only 16.3% of hits.
54   My *Military Thought in the French Army, 1815–51* (Manchester University 1989)
pp. 122–5. Cf many of the un-standardized skirmisher drill books and schemes of the
revolutionary period can be found in AHG MR 2034, 2041, 2043, 1962, 2008, 2012, Xs
143. See also Duhesme, op. cit., and La Roche Aymon *Des Troupes Légères* (Paris, 1817).
55   Houssaye 1814, and my *A Book of Sandhurst Wargames* (Hutchinson, London 1982)
pp. 25–30.
56   This emerges in e.g. Liddell Hart, B. H., ed., *The Letters of Private Wheeler*
(Joseph, London 1951).
57   Weller op. cit. Gates, op. cit., pp. 79, 144–5, suggests that in battle there could not
be great accuracy with rifles at much more than 150 yards, although on pp. 96, 144, he
shows that Rottenberg's training manual envisaged target practice up to 300 yards.
58   Baker, E., *Remarks on the Rifle* (first published Brighton, 1805; 11th edn., London
1835, reprinted Standard publications, Huntington, West Virginia n.d., c. 1960?), p. 53.
Note that his rifle was sighted to 200 yards.
59   Marshall, S. L. A., *Infantry Operations and Weapons Usage in Korea* (E. C. Ezell ed.,
Greenhill, London 1988) pp. 7–8. See also Chapter 5, below, for Vietnam.
60   Busk, op. cit., pp. 22–3; Haythornthwaite, op. cit., pp. 18–19; and Gates, op. cit.,
p. 142, for rates of fire.
61   For Wellington at Toulouse see Oman, *Peninsular War*, op. cit., vol. VII, p. 475; for
the Imperial Guard at Essling see Lachouque, H., *The Anatomy of Glory* (trans. A. S. K.
Brown, new edn., Brown University, Rhode Island 1962), p. 156.
62   At Maya an entire British battalion was formed on a frontage of 25 yards, with steep
slopes on either side, making perhaps 20 or 30 men per yard – although it would only be
the first two ranks that could use their muskets effectively. Wellington at Waterloo had
almost 21,000 men in his front line, which was a mile and a half long – as many as eight
men per yard – while in more normal battles the effective density in the front line rarely
fell lower than 5 men per yard.
63   For the musket found at Gettysburg loaded 23 times, see my *Rally Once Again*, op.
cit., p. 86.
64   For Salamanca see Glover, *Wellington's Peninsular Victories*, p. 83; for Toulouse see
Oman, *Peninsular War*, op. cit., vol. VII, p. 476.
65   Bressonet, op. cit., p. 135.
66   Oman, *Peninsular War*, op. cit., vol. VII, p. 360.
67   Norris, A. H., and Bremner, R. W., *The Lines of Torres Vedras* (London 1980)
pp. 14 *ff.*
68   Duffy, C. J., *Borodino, Napoleon Against Russia 1812* (Sphere edn., London 1972)
pp. 102, 125–9.
69   Beatson, F. C., *With Wellington in the Pyrenees* (Goschen, London n.d., 1914?).
70   Gleig, G. R., *The Subaltern*, p. 112 describes the cantonment of 100 men in a
farmhouse, and on p. 151 the whole regiment in a cottage.
71   New Orleans is covered in Ward, J. W., *Andrew Jackson, Symbol for an Age* (Oxford
University Press, New York 1962); Brooks, C. B., *The Siege of New Orleans* (University
of Washington, Seattle 1961); Reid, J. and Eaton, J. H., *The Life of Andrew Jackson*
(F. L. Owsley jr, ed., University of Alabama 1974); Carter, S., 3rd, *Blaze of Glory*
(St Martins, New York 1971); Reilly, R., *The British at the Gates* (Cassell, London 1974);
James, W., *Military Occurrences Between Great Britain and America* (2 vols., London,
1818), vol. II; Casey, P., *Louisiana in the War of 1812* (privately printed, Baton Rouge
1963); and Surtees, W., *Twenty-Five Years in the Rifle Brigade* (first published 1833,
Military Book Society reprint, London 1973).
72   Ward, op. cit., p. 4. It was arguably in the British interest to lift this American
gloom, to pre-empt a possible renunciation of the treaty!
73   It was on the west bank that most of the famous 'Hunters of Kentucky' were actually

posted, not in the main battle. When finally attacked by Thornton's 450 men later in the morning, over a thousand of the Kentuckians ran away after receiving a volley at 150 yards. They inflicted but 33 casualties on the British, and abandoned their colours and nine cannon. See Brooks, op. cit., p. 242, and Carter, op. cit., p. 237.

74 Burgoyne, R. H., *Historical Records of the 93rd Sutherland Highlanders* (London, Bentley, 1883) pp. 29–34, 36–45.

75 Carter, op. cit., pp. 247, 253.

76 Among the ranges variously quoted for the opening of fire are 150 yards (Reilly p. 299); 200 yards (James vol. II, p. 545), and 250 yards (Casey, p. 84).

77 Carter, op. cit., p. 254; cf Reid and Eaton, op. cit., p. 339, say all the Americans opened fire at once, and at quite short range because of the mist.

78 Casey, op. cit., pp. 73, 84; Surtees, op. cit., p. 376; and James vol. II, op. cit., p. 381. On p. 75 Casey states that the Kentucky troops involved in these actions were firing buckshot, not using rifles. See Ward, op. cit., pp. 16–26, for another interesting corrective to the 'frontier rifleman' myth at New Orleans.

79 Surtees, op. cit., pp. 373–4.

80 Surtees, op. cit., pp. 375–6.

81 Burgoyne, op. cit., p. 33.

82 Surtees, op. cit., pp. 389–90.

Chapter 3

1 E.g. Porch, D., 'The French Army and the Spirit of the Offensive 1900–14', in Bond, B., and Roy, I., *War and Society – A Yearbook of Military History* (London, 1975), p. 117; Travers, T.H., *The Killing Ground* (Allen & Unwin, London 1987); or Ellis, J., *Eye-deep in Hell. The Western Front 1914–18* (London, 1976), Chapter 6, 'Strategy and Tactics'.

2 The earliest use of the term 'firefight' which I have come across was in February 1853 in the *Military Review* II, p. 1. Quoted in Strachan, H., 'Wellington's Legacy: The British Army in the Age of Reform' (unpublished thesis, Cambridge University, 1978).

3 *Waterloo Letters*, op. cit., pp. 365–6.

4 Kincaid, J., *Adventures in the Rifle Brigade, in the Peninsula, France and the Netherlands from 1809 to 1815*, ed. Fortescue, J. (London, 1929), p. 255.

5 Clausewitz, C. von, *On War*, translated by Graham, J. J. (new edition 3 vols., London, 1940), vol. 3, pp. 263–4.

6 Hofschroer, P., *Prussian Light Infantry, 1792–1815* (Osprey, London 1984), pp. 10–27, and his *Prussian Line Infantry, 1792–1815* (Osprey, London 1984), pp. 10–16. For the French *Grandes Bandes*, see J. Arnold, op. cit.

7 Lauerma, M., *L'Artillerie de Campagne Francaise Pendant les Guerres de la Révolution* (Helsinki, 1956). For a British example of a battlefield made empty by two contending artilleries, we need look no further than the two opening rounds at New Orleans, on 28 December 1814 and 1 January 1815.

8 Duc d'Orléans' cyclostyled letter to the Minister of War, 1840, on the Chasseurs, p. 12. In AHG MR 1947.

9 Fuller, J. F. C., *Sir John Moore's System of Training* (London, 1924). Sir John Moore's personal role in training the Light Division has recently been called into question: see Gates, op. cit., p. 112.

10 Paret, P., *Yorck and the Era of Prussian Reform, 1807–1815* (Princeton, 1966).

11 Blackmore, *British Military Firearms 1650–1850* (London, 1961), pp. 85–7.

12 Barré MSS on percussion muskets, December 1830, in AHG MR 2141, *Projects, 1830–33 (A) – Machines*. For a similar American experience with the Hall rifle, from

1811 onwards, see Davis, C. L., *Arming the Union* (National University, New York 1973), pp. 107–14.

13   Morand, C. A. L. A., *De l'Armée Selon la Charte* (Paris, 1829), p. 235.

14   Orléans on Chasseurs, op. cit., p. 16; Col. Marnier, cyclostyled article 'Améliorations Proposées dans l'Armement et l'Education des Troupes', May 1837, pp. 19–20, in AHG MR 2140; also Ardant du Picq MSS 'De l'emploi de la Carabine et des Chasseurs' n.d., p. 1, in AHG MR 1990.

15   Hamilton, op. cit., vol. 1, p. 144.

16   Le Louterel, 'Essai de Conferences sur l'emploi des Manoeuvres d'Infanterie devant l'Ennemi' (Paris, 1848), p. 11.

17   Bapst, C. G., *Le Marechal Canrobert* (6 vols., Paris, 1898–1913), vol. 1, p. 362. For all these debates see my *Military Thought in the French Army, 1815–51* (Manchester University Press, 1989).

18   Engels, F., *Engels as a Military Critic* (eds. Chaloner and Henderson, London, 1959), p. 79 ff.

19   MSS 27th April 1844, in AHG Xs 141.

20   MSS 10th September 1833, 'Quelques Observations sur la Formation de l'Infanterie sur Deux ou Trois Rangs', in AHG MR 2012.

21   Delorme du Quesney, A., *Du Tir des Armes à Feu* (Paris, 1845), p. 176.

22   Bugeaud, T. R., *Apercus sur Quelques Details de la Guerre* (First published in 1832. 24th edn. Paris, 1873).

23   Ibid., p. 145.

24   See my article 'The Strategic Challenge to the French Engineers, 1815–51', in *Fort* (the Journal of the Fortress Study Group), no. 5, Spring 1978, p. 31.

25   See p. 132 of my thesis (op. cit.) and Spivak, M., 'Le Colonel Armoros, un Promoteur de l'Education Physique dans l'Armée Francaise', in *Revue Historique de L'Armée*, 1970.

26   Thoumas, C. A., *Les Transformations de l'Armée Francaise* (2 vols., Paris, 1887), vol. II, p. 452.

27   Luvaas, J., *The Military Legacy of the Civil War – The European Inheritance* (Chicago, 1959), pp. 150 *et seq.*

28   Lecomte, F., *Relation Historique et Critique de la Campagne d'Italie en 1859* (Paris, 1860), p. 126.

29   Ibid. For the Italian campaign see also Wylly, H.C., *1859, Magenta and Solferino* (Swan Sonnenschein, London 1907).

30   Ardant du Picq, op. cit., p. 102.

31   Ibid., p. 128.

32   Ibid., p. 147.

33   Ibid., p. 116.

34   For France, 1867–84, see Thoumas, op. cit., vol. II, pp. 456–64; and Holmes, E. R., 'The Road to Sedan, the French Army 1866–70' (Royal Historical Society, London 1984). For Britain, 1902–9, see Travers, T. H. E., 'The Offensive and the Problem of Innovation in British Military Thought 1870–1915' in *Journal of Contemporary History*, vol. 13, no. 3, July 1978, p. 531.

35   Bonnal, H., *Sadowa, a Study*, trans. Atkinson, C. F. (2nd. Imp., London, 1913), p. 237.

36   Prince Kraft zu Hohenlohe Ingelfingen, *Letters on Infantry*, trans. Walford, N. L. (London, 1889), pp. 135–6. See also Wilhelm Duke of Württemberg, *The System of Attack of the Prussian Infantry in the Campaign of 1870–71* (Trans. C. W. Robinson, Aldershot 1871).

37   Herbert, W. V., *The Defence of Plevna 1877* (London, 1895), p. 281.

38   Forbes, Macgahan *et al.*, *The War Correspondence of the 'Daily News', 1877* (3rd edn., London, 1878), p. 365.

39 Herbert, op. cit., p. 198.
40 Hamilton, op. cit., vol. 2, pp. 203–6.
41 Ibid., vol. 2, p. 309.
42 Rommel, E., *Infantry Attacks*, trans. Kiddé, G. E. (US Marine Corps Association, Quantico, Virginia, 1956), p. 7.
43 Ibid., pp. 10–11.
44 Ibid., p. 16.
45 From *War and Policy* (New York, 1900), p. 159, quoted in a different sense in Ellis, J., *The Social History of the Machine Gun* (London, 1975), p. 53.
46 Hamilton, op. cit., vol. I, p. 114.
47 Bloch, I. S., *Modern Weapons and Modern War*, ed. Stead, W. T. (London, 1899).
48 Tactical lessons drawn from the Boer War by the British have been extensively analysed in recent years in Bidwell, S. and Graham, D., *Firepower* (Allen & Unwin, London, 1982); Travers, T., *The Killing Ground* (Allen & Unwin, London, 1987); and Badsey, S. D., *Fire and the Sword* (unpublished doctoral thesis, Cambridge, 1981, due to be published by Manchester University Press).
49 Actually it was won by Maxim guns, with the revealing final 'score' of around 400 British casualties to 30,000 Dervishes. Ironically, this was to be about the same proportion of casualties, albeit on a smaller scale, as was suffered between the Germans and the British on the First Day of the Somme.
50 Esposito, V. J., ed., *The West Point Atlas of American Wars* (2 vols., Praeger, New York 1959), vol. I, *1689–1900*.
51 Dupuy, T. N., *The Evolution of Weapons and Warfare* (first published 1980; Jane's edn., London 1982) p. 191. The latest scholarly re-working of this time-honoured theme is Hagerman, E., *The American Civil War and the Origins of Modern Warfare* (Indiana University, 1988); but see also Mahon J. K., *Civil War Assault Tactics* in *Military Affairs*, vol. 25, summer 1965, pp. 57–68; and McWhiney, G., and Jamieson, P. D., *Attack and Die* (University of Alabama, 1982).
52 My book *Rally Once Again* (Crowood, 1987, published in USA by Yale University as *Battle Tactics of the Civil War*, 1989) devotes some 230 pages to explaining this quite simple point. Unless otherwise indicated in the footnotes, it is the reference for the present section.
53 Wynne, W. R. C., *Memoir* (privately printed by the family, Southampton c. 1880; to be re-published by Fieldbooks, Camberley 1990, edited by Howard Whitehouse), p. 136.
54 See especially *Rally Once Again*, op. cit., pp. 88–9, 145–50. For a parallel that would be ludicrous if it were not true, the reader is referred to the gunfight at the OK Corral, 1881, where the engagement range was apparently between six and eight feet.
55 Hardee, W. J., *Rifle and Light Infantry Tactics for the Exercise and Manoeuvres of Troops* (2 vols, first published 1855; reprinted Greenwood, Westport, Conn. 1971).
56 *Rally Once Again*, op. cit., pp. 150–4.
57 Johnson, R. U. and Buel, C. C., eds., *Battles and Leaders of the Civil War* (4 vols., Century, New York 1884 and many later reprints), vol. II, p. 510.
58 Ibid., pp. 29–72 and Appendix I. One would certainly not wish to suggest, however, that they were anything but fully committed to charges which were unavoidable.
59 Ibid., pp. 39–40, 50–51, 133–5.
60 Ashworth, *Trench Warfare 1914–1918: The Live and Let Live System* (Macmillan, London 1980).
61 *Rally Once Again*, op. cit., pp. 154–8.
62 *A New System of Infantry Tactics, Double and Single Rank, adapted to American Topography and Improved Firearms* (Appleton, New York, 1867).
63 Weigley, R. F., in Paret, *Makers of Modern Strategy*, op. cit., pp. 413–18; and *Rally Once Again*, op. cit., pp. 123–35. H. W. Halleck's book was *Elements of Military Art and*

Science (Appleton, New York 1846, reprinted by Greenwood, Westport, Conn. 1971).

64 Hagerman, op. cit., has the most thorough analysis of the rise of fortifications.

65 E.g. Dupuy, p. 171. For my analysis of similar claims, see *Rally Once Again*, op. cit., pp. 165–78.

66 Against rifled artillery not even the best infantry weapons could outrange the guns. In the field, however, rifled cannon fired relatively small and technically less effective rounds than the 12-pounder smoothbores. Gunners normally wanted to deploy two smoothbores to every one rifle.

67 *Rally Once Again*, op. cit., pp. 179–88, and see Badsey, op. cit., pp. 15, 21–7.

68 *Moulin à café* fire is in Württemberg, op. cit., pp. 8–10. Badsey, op. cit., pp. 15, 30, describes problems encountered with the needle gun's gas seal, requiring fire from the hip to avoid back blast. Note that J. A. English agrees that the important mid-century change came with the needle gun, not the rifle musket – *On Infantry* (first published Praeger, New York 1981; new edn 1984), p. 1.

69 E.g. Howard, M. E., *The Franco Prussian War* (first published 1960, Fontana edn., London 1967), p. 36.

70 Ibid., pp. 215–6.

71 Ascoli, D., *A Day of Battle* (Harrap, London 1987) pp. 168–72; Captain Loir, *Cavalerie* (Paris, Chapelot 1912) pp. 308–317; and elsewhere in Loir for other effective uses of cavalry in 1870. Howard, however, scoffs: op. cit., p. 157.

72 Morris, D., *The Washing of the Spears* (first published 1965, Sphere edn., London 1968), pp. 300, 371–6; see also the diagram in Whitehouse, H., *Battle in Africa* (Fieldbooks, Camberley 1987) p. 35.

73 Barrow, E. G., *Infantry Fire Tactics* (Hong Kong, 1895, reprinted by Wargame Library, Hemel Hempstead, Herts, 1985), pp. 4 ff. There was a story that at Omdurman troops armed with the Martini-Henry held the enemy attack at 300 yards; those with the Lee Metford at 500.

74 De Grandmaison, L., *Dressage de l'Infanterie en Vue du Combat Offensif* (Berger-Levrault, Paris 1906) p. 6; de Maud'huy, L., *Infanterie* (Lavauzelle, Paris 1911) p. 103.

75 Balck, W., *Tactics* (new edn., Posen 1908; trans. W. Kreuger, Fort Leavenworth 1911; reprinted Greenwood Press, Westport Conn., 1977), vol. I, pp. 137, 150–3, 162.

76 The title is taken from graffiti chalked on trains taking troops to the front in August 1914, '*Train de Plaisir pour Berlin*' – see e.g. Blond, G., *La Marne* (new edn., Livre du Poche, Paris 1962), p. 7.

77 Much of the technical background is in Porch, D., *The March to the Marne, The French Army 1871–1914* (Cambridge University, 1981). The weapons are listed in Hicks, J. E., *French Military Weapons, 1717–1939* (Flayderman, New Milford, Conn. 1964).

78 A useful general military history is Contamine, H., *La Revanche, 1871–1914* (Berger-Levrault, Paris 1957); see also Hanotaux, G., *L'Enigme de Charleroi* (Edition Française Illustrée, Paris 1917). The quotation comes from a telling prediction in Captain 'Danrit', *La Guerre de Demain* (6 vols., Flammarion, Paris 1891), Part I, *La Guerre en Rase Campagne*, vol. I, p. 320.

79 The justice of the French view of the BEF leaps out from between the lines of the British official history, *Military Operations, France and Belgium 1914* (Macmillan, London 1923) vol. I; and Terraine, J., *Mons, the Retreat to Victory* (Batsford, 1960). Essential reading in this context is General Lanrezac's *Le Plan de Campagne Français et le Premier Mois de la Guerre* (Payot, Paris 1920).

80 The ideological background may be found in three fascinating and complementary books – Sternhell, Z., *La Droite Révolutionnaire, 1885–1914* (Seuil, Paris 1978); Girardet, R., *L'Idée Coloniale en France de 1871 à 1962* (La Table Ronde, Paris 1972); and Digeon, C., *La Crise Allemande de la Pensée Française, 1870–1914* (PUF, Paris 1959).

81 Actually an inspection of the ground today shows that this 'battlefield' is a shell trap,

badly chosen for the material aspects of defence, whatever may have been its spiritual advantages. The battle is described in *La Guerre de Demain*, op. cit., vol. I, pp. 332 ff, of which the pseudonymous author was in fact Lt. Col. E. Driant, an early Fascist agitator celebrated since 1916 as a significant Verdun martyr.

82   Dragomirov, P., *Les Étapes de Jeanne d'Arc* in *Revue des Deux Mondes*, 1 March 1898, pp. 153–76. Péguy, C., *Oeuvres Complètes* (Nouvelle Revue, Paris 1933). Péguy was killed in action in 1914.

A modern view of Joan is Warner, M., *Joan of Arc* (Penguin, London 1983).

83   The evolution of French planning is in Contamine, op. cit., p. 59 ff; Gambier, F., and Suire, M., *Histoire de la Première Guerre Mondiale* (2 vols., Fayard, Paris 1968), vol. I, pp. 157–75; and the French official history, *Les Armées Françaises dans la Grande Guerre* (Ministère de la Guerre, Paris 1922), book I, vol. I, pp. 1–36.

84   Balck, W., *Tactics*, op. cit., vol. I, p. 119n, shows that in French tests red (and presumably also dark blue) came towards the middle of the spectrum of inconspicuous colours in the field. Greens and browns were better, but light blue was worse.

85   Military politics are in Porch, op. cit., and de la Gorce, P.-M., *The French Army* (trans. K. Douglas, Weidenfeld & Nicolson, London 1963), pp. 17–92. The 1905 recruitment law reduced the term of service to two years. This was increased to three in 1913 but by then, of course, it was a little late . . .

86   Danrit, op. cit., vol. I, p. 378; cf. he strongly recommends the bayonet counter-charge in vol. II, p. 31.

87   The story is traced in Contamine, op. cit., *passim*; Carrias, *La Pensée Militaire Française* (PUF, Paris 1960) pp. 268–301; and Possony, S. T., and Mantoux, E., *Du Picq and Foch: The French School* in Earle, E. M., ed., *Makers of Modern Strategy* (Princeton University, 1943) Chapter 9, pp. 206–32.

88   *Règlement de Manoeuvre d'Infanterie du 20 Avril 1914* (Chapelot, Paris 1914).

89   Ibid., p. 14. Compare Holmes, op. cit., for a similar confusion of tactics in 1868–70.

90   *Règlement de Manoeuvre*, op. cit., p. 69.

91   Telling passages are in ibid., pp. 23, 77, 115–23, 140–1.

92   *Psychologie des Foules* (PUF, Paris 1947 edn). The impact of this book is extensively discussed in Nye, R. A., *The Origins of Crowd Psychology* (SAGE, London 1975), Chapter 6, pp. 123–53, *Gustave Le Bon and Crowd Theory in French Military Thought Prior to the First World War*.

93   *Infanterie*, op. cit., pp. 14–29. Fatigue and nervous exhaustion are treated at greater length on pp. 47–87. Compare the equally perceptive analysis of combat shock and leadership in General Percin's *Le Combat* (2nd edn., Alcan, Paris 1914), Chapter 4, pp. 34–84.

94   De Maud'huy, op. cit., p. 33.

95   Ibid., pp. 30–46, 97–100.

96   De Grandmaison, published by Berger-Levrault, Paris.

97   Ibid., pp. 2–3, 20–2, 42–7.

98   *Deux conférences faites aux officiers de l'état-major de l'armée: La notion de sûreté, et l'engagement des grandes unités* (Berger-Levrault, Paris 1911).

99   Ibid., pp. vi, 1–4.

100   Ibid., p. 40.

101   This problem continued to plague French exercises even after de Grandmaison's work had been published; see e.g. General Palat's *Les Manoeuvres de Languedoc en 1913* in *Revue des Deux Mondes*, 15 October 1913, pp. 799–817; and Percin, op. cit., pp. 152–3.

102   Gambiez, F., and Suire, M., op. cit., vol. I, pp. 108 ff, accept that 'de Grandmaison himself was in practice far less crazy than his explosive phrases might suggest'.

103   The full story of Joffre's thinking has never been properly told, but see e.g. *Mémoires du Maréchal Joffre, 1910–17* (2 vols., Plon, Paris 1932) vol. I, pp. 224–94.

104   Lanrezac, op. cit., p. 150–95.
105   General de Castelli, *Le Sème. Corps en Lorraine, Août–Octobre 1914* (Berger-Levrault, Paris 1925), pp. 27–58. See also French official history, op. cit., book I, vol. I, pp. 200–10.
106   Contamine, op. cit., pp. 244–5, for de Grandmaison. Colin, Gen H., *Les Gars du 26ème* (Payot, Paris 1932), pp. 50–5, for the actions around Conthil.
107   Contamine, op. cit., pp. 238–9. A detailed eyewitness account from the 143rd regiment, in front of Loudrefing, implies that at least half of its approximately 1,000 casualties were suffered while in a defensive posture (records of the 32nd Division in AHG carton 25 N 147: Anon. ms. report, *Opérations des 19 et 20 Août*).
108   French official history, op. cit., book I, vol. I, pp. 245–59.
109   General Veron, *Souvenirs de ma Vie Militaire – Impressions et Réflexions* (Maugard, Rouen 1969), pp. 57–8.
110   For artillery, see e.g. Porch op. cit., pp. 232–7, and Percin, op. cit., pp. 165–208.

Chapter 4

1   Wintringham & Blashford-Snell, op. cit., p. 167.
2   De Gaulle, C., *The Army of the Future* (trans. London, 1940), p. 42.
3   Jünger, E., *The Storm of Steel* (Mottram, R. H., ed., London, 1929), p. *vi*.
4   Ibid., pp. 107–8.
5   Ibid., p. 235.
6   Ibid., p. 110.
7   Rommel, E., op. cit., p. 35.
8   Wynne, G. C., *If Germany Attacks* (London, 1940); see also his article 'Pattern for Limited War; the Riddle of the Schlieffen Plan', 3 parts, in *Royal United Services Institution Journal*, 1958–9.
9   Jünger, op. cit., pp. 286–7. For the uselessness of the tank compare Terraine, J., *The Smoke and the Fire* (London, 1980), pp. 148–60.
10   Ibid., p. 301.
11   Quoted in 'Report on the Staff Conference held at the Staff College, Camberley, 9–11th January, 1933' (PRO WO 32 (3116)), p. 30.
12   Wintringham, T., *English Captain* (London, 1939), p. 304.
13   Ibid., p. 306.
14   De Gaulle, op. cit., p. 47.
15   Ibid., p. 103.
16   Major General McNamara, in 'Report on the Staff Conference', &c., op. cit., p. 16.
17   Quoted in Trythall, A. J., *'Boney' Fuller; the Intellectual General* (London, 1977), p. 82.
18   Ibid., p. 27. The most recent general treatment of Fuller's work is Reid, B. H., *J. F. C. Fuller, Military Thinker* (Macmillan, London 1987).
19   Guderian, H., *Panzer Leader* (trans. Fitzgibbon, C., Futura edn., London, 1979), p. 21.
20   Ibid., p. 24.
21   Ibid., p. 106.
22   Chuikov, V. I., *The End of the Third Reich* (trans. Kisch, R., Panther edn., London, 1969), p. 60.
23   See also Ellis, J., *The Sharp End of War* (London, 1980), Chapter 3, 'Combat: Infantry', pp. 52–116.
24   See tactical diagrams in Bidwell and Graham, op. cit., pp. 236–7. 'Retreating

through your anti-tank screen' was a gambit suggested by Fuller before the war – Reid, op. cit., p. 166.

25 Schmidt, H. W., *With Rommel in the Desert* (London, 1952), p. 78.

26 For the 'Snipe' action I have relied totally upon Lucas Phillips, C. E., *Alamein* (London, 1962), pp. 262–302.

27 Ibid., p. 296.

28 Orgill, D., *The Gothic Line* (London, 1967), pp. 112–13.

29 Quoted in Ibid., pp. 95–6, from Bright, J., ed., *The 9th Queen's Royal Lancers 1936–45*, pp. 168–9.

30 Bishop, G. S. C., *The Battle – a Tank Officer Remembers* (Fotodirect Printers, Brighton, Sussex, *n.d.*), p. 60.

31 Ibid., p. 78.

32 Marshall, S. L. A., *Men Against Fire* (New York, 1947), pp. 44–5. Spiller, R. J., *S. L. A. Marshall and the Ratio of Fire*, in *JRUSI*, vol. 133, no. 4, winter 1988, pp. 63–71, shows that Marshall did not conduct nearly as many systematic post-combat interviews as he often claimed. His contribution was less scientific, and more like the good, intuitive press reporter that he was at heart.

33 Ibid., p. 89.

34 Lindsay, M., *So Few Got Through* (London, 1946), p. 247.

35 Ibid., p. 128.

36 Ibid., p. 216.

37 For battleshock see *e.g.* Lord Moran, *The Anatomy of Courage* (first published 1985, new edn. Constable, London 1966); Dinter, E., *Hero or Coward* (Cass, London 1985); Ellis, J., *Sharp End*, op. cit.

38 Marshall, op. cit., p. 19.

39 Ibid., pp. 208–9.

40 Badsey, op. cit., pp. 283–343.

41 Badsey, op. cit., especially pp. 305, 320, 334–5, 342–3. The general non-use of a potentially effective battle cavalry in 1914–18 shows many similarities with the same phenomenon in the American Civil War.

42 For the British, see Bidwell & Graham, op. cit., pp. 61–146, especially pp. 122–9. For the French, Guinard, P., Devos, J.-C., and Nicot, J., *Inventaire Sommaire des Archives de la Guerre, Série N 1872–1919* (Renaissance, Troyes 1975), pp. 123–31.

43 For artillery, see Bidwell and Graham, op. cit., especially pp. 94–114, and pp. 140–3 for signals. For a sociological explanation of why the continuity of the 'psychological' or 'human' battlefield died hard, Travers, *The Killing Ground*, op. cit., pp. 37–82, 250–3; and C. S. Forester's *The General*, for a fictionalised account of how it did (allegedly based on Allenby).

44 Champions of the Western Front and its generals liked to portray their heroes manfully surmounting the challenge of radically novel difficulties, while the anti-militarists almost revelled in the no less unprecedented quality of the horror and futility. Apologists such as Cyril Falls attempted to correct the negative tone of 'war books' (see his own war book of that name, Davies, London 1930), and in recent times the same mantle has fallen to John Terraine; but by and large they have failed to establish the continuity of warfare in the popular mind.

45 The best air futurism was in Giulio Douhet's *The Command of the Air* (first published 1921, English edn. trans. D. Ferrari, Coward McCann, New York 1942). A handy short technical history of bombing in World War II is Frankland, N., *The Bombing Offensive Against Germany* (Faber, London 1965). The economic ineffectiveness of the campaign is explained in Milward, A. S., *The German Economy at War* (Athlone, London 1965).

46 Harris, J. P., *The British General Staff and the Coming of War, 1933–9*, in *Bulletin of the Institute of Historical Research*, vol. LIX, no. 140, November 1986, pp. 196–211.

47 Basil Liddell Hart was a British infantry officer, gassed on the Somme, who helped

General Maxse reform infantry tactics towards the end of the war and later publicised himself by recommending a (rather vague) 'indirect approach' using first armour and then, from the mid-1930s, strategic terror-bombing instead. In the late 1930s he did much to damage the cause of armour in Britain, using a highly selective and self serving interpretation of the past. It was not accidentally that his various writings were omitted from the first edition of this book, and it is gratifying to report that J. J. Mearsheimer's meticulous *Liddell Hart and the Weight of History* (Cornell University, New York, 1988) has now at last placed this insecure journalist and his academic devotees in a true perspective.

48 Harris, J. P., *British Armour and Rearmament in the 1930s*, in *The Journal of Strategic Studies*, vol. 11, no. 2, June 1988, pp. 220–44. Note that the Germans at this time had to be content with equally light tanks – but by 1943 they had solved the problem by fielding the Panther and Tiger.

49 General Wilson said of Hobart that 'his tactical ideas are based on the invincibility and invulnerability of the tank to the exclusion of the employment of other arms in correct proportion'. Quoted in Macksey, K., *Armoured Crusader* (Hutchinson, London 1967) p. 170.

50 E.g. *The Armoured Regiment* (War Office, London, July 1940); and *Handling of an Armoured Division* (War Office, London, May 1941). Despite his later attempts to distance himself from Fuller's view of the proportion of infantry to accompany tanks, Liddell Hart actually shared it. Neither of them wanted infantry to have a very great role, but both wanted it to have some: Reid, op. cit., pp. 159–64.

51 Reid, op. cit., pp. 49, 165–6.

52 German weakness in tank strength, 1940, is in *e.g.* Macksey, K., *Tank Warfare* (first published 1971, Panther edn., London 1976), pp. 110–12. Carver's significant study of their relatively inferior technology is summarised in Agar-Hamilton, J. A. I., and Turner, L. C. F., *The Sidi Rezeg Battles, 1941* (Oxford University Press, Cape Town 1957), pp. 31–44; although it must be admitted that their tanks were more reliable and all had radios.

53 Macksey, K., *Beda Fomm, the classic victory* (Ballantine, London 1972), pp. 135–51.

54 See *e.g.* Rommel's comments reported by von Mellenthin, F. W., *Panzer Battles* (trans. Betzler, Cassell, London 1955 and Oklahoma University 1956), p. 63.

55 Armoured divisions were Ariete, 15th and 21st Panzer: motor divisions were Trieste and 90th Light: infantry divisions included Pavia, Savona, Bologna, Brescia. They enjoyed a considerable continuity under Rommel's command.

56 Note that by the time of the Normandy campaign the German infantry was stiffened by (defensive) lightly armoured self-propelled anti-tank guns, and (offensive) heavier armoured assault guns – i.e. turretless tanks which could fill most of the roles of the British 'I' tanks.

57 In theory the tank battalions had 53 tanks each, but very often the entire division might have only 20–30 tanks running at a time.

58 Bidwell and Graham, op. cit., pp. 234–5; Agar-Hamilton and Turner, op. cit., pp. 56–8.

59 Orders of battle in Agar-Hamilton and Turner, op. cit., pp. 474–5.

60 'I' (or 'Infantry') tanks were intended to be heavily armoured, but not necessarily fast-moving or heavily gunned.

61 *E.g.* Agar-Hamilton, op. cit., p. 65.

62 In *Alamein to the River Sangro* (first published 1948, Grey Arrow edn. 1960), p. 36, Montgomery said 'methodical', 'crumbling' action is 'within the capabilities of my troops' (i.e. infantry troops).

63 Carver, M., *Dilemmas of the Desert War* (Batsford, London 1986), pp. 50–3.

64 Before 'Crusader' Auchinleck had wanted each armoured division to have one tank

and one infantry brigade, 'like a Panzer Division'; but in the event this reorganisation took over a year to implement.

65 Bidwell and Graham, op. cit., pp. 222, 245; Hamilton, N., *Montgomery* (3 vols., Hamish Hamilton, London 1981–6), vol. 1, *The Making of a General*, pp. 611–13, 617–20, 630.

66 For the parrot compare Farrar-Hockley, op. cit., p. 8, with Lindsay, op. cit., p. 35. For Montgomery's desire to seize the initiative, see *e.g.* Hamilton, op. cit., vol. I, p. 590.

67 The idea was to group two or more armoured divisions together as a mobile reserve, 'like in the Afrika Korps'. The conception and creation of this force – officially called X Armoured Corps – is recounted in Hamilton, op. cit., vol. I, pp. 589, 591, 641–2. Its novelty is denied in Correlli Barnett, *The Desert Generals* (Kimber, London 1960) pp. 252–3.

68 Badsey, op. cit., p. 302.

69 Ibid., pp. 302–9.

70 Playfair, I. S. O., and Moloney, C. J. C., *The Mediterranean and Middle East* (British official history, HMSO, London 1966), vol. IV, p. 77.

71 Hamilton, op. cit., vol. I, pp. 815–24.

72 D'Este, C., *Decision in Normandy* (Pan edn., London 1984), p. 356.

73 Ibid., pp. 290, 352–69.

74 Field Marshal Carver, quoted in Ibid., p. 290.

75 Ibid., pp. 400–7.

76 Playfair and Moloney, op. cit., vol. IV, pp. 34–5.

77 Both battles are in Horrocks, *Corps Commander* (first published 1977, Magnum edn., London 1979) pp. 82–111, 151–68. The desired Reichswald timings seem too embarrassing to be specifically mentioned in histories and memoirs, but there are strong hints in the operation orders reprinted in *Battlefield Tour – Operation Veritable* (BAOR, Germany, December 1947), pp. 27–31. See also Elstob, P., *Battle of the Reichswald* (Ballantine, London 1970) pp. 66, 75.

78 D'Este, op. cit., p. 65.

79 Ibid., pp. 79–81.

80 His takeover of command in the desert is in Hamilton, op. cit., vol. I, pp. 606–36, while preparations for Alam Halfa are on pp. 637–48, 661–8.

81 Montgomery, *El Alamein to the River Sangro*, op. cit., p. 18.

82 'His' own account (actually written by a ghost writer) in *El Alamein to the River Sangro*, op. cit., and *Normandy to the Baltic* (Hutchinson, London 1946) insists that nothing ever wavered from the plan, with the classic case being the Normandy campaign in general and the battles around Caen in particular. D'Este's book, op. cit., is devoted largely to exploding this claim for Normandy.

83 A recent example is Hastings, M., *Overlord: D Day and the Battle for Normandy* (Joseph, London 1984). Creveld, M., *Fighting Power* (Arms & Armour, London 1983) compares US and German practice, concluding that the Germans were superior because they designed their military institutions around the needs of combat, and ensured their best men were in the front line.

84 Jary, S., *18 Platoon* (Sydney Jary, Carshalton Beeches, Surrey, 1987) p. 21.

85 Ibid., p. 17, although Jary did miss his battalion's most intense fighting in Normandy. Other voices raised against military teutomania include Peters, R., *The Dangerous Romance; the US army's fascination with the Wehrmacht* in *Military Intelligence*, October–December 1986, pp. 45–8; Beaumont, R. A. *On the Wehrmacht Mystique* in *Military Review*; vol. 66, no. 7, July 1986, pp. 44–56; Hughes, D. J., *Abuses of German Military History* in *Military Review*, vol. 66, no. 12, December 1986, pp. 67–76. Creveld himself replied to the last two in *On Learning from the Wehrmacht* in *Military Review*; vol. 68, no. 1, January 1988, pp. 63–71.

86 American defence statisticians have calculated that in June 1943 one German had the combat effectiveness of just over three Russians, although by 1944 this had fallen to 1.6 Russians (as compared to 1.2 British or Americans): Dunnigan, J. F., ed., *The Russian Front* (Arms & Armour, London 1978) pp. 82–3; and related discussion in Creveld, *Fighting Power*, op. cit., pp. 5–10.

87 Ms letter of Balck to Capt. R. d'A. R. Ryan, 18 August 1960, privately communicated to the author.

88 Balck's actions on the Chir are in Mellenthin, op. cit., pp. 175–84. Other stirring memoirs include Guderian, op. cit.; E. von Manstein, *Lost Victories* (first published 1955, translated Methuen, London 1958); Anon, *German Defense Tactics Against Russian Break – Throughs* (US Army Department 20–233, Washington DC, October 1951). Another classic, albeit sadly inaccessible to the present author, is von Senger und Etterlin, F. M., *Der Gegenschlag* (Neckargemünd, 1959).

89 Dunnigan, op. cit., pp. 8, 15, 22, 97, 117–8, 142.

90 Simpkin, R., *Race to the Swift* (Brassey, London 1985) pp. 17, 37–9, 46.

91 This, and much of what follows, is drawn from Dick, C. J., *The Operational Employment of Soviet Armour in the Great Patriotic War*, in Harris, J. P., and Toase, F. H., eds., *Modern Armoured Warfare* (forthcoming from Batsford, London 1990); and Dunnigan, op. cit., pp. 98–106, Chapter 4, *Organization of Soviet Ground Forces*.

92 Bellamy, C., *Red God of War* (Brassey, London 1986) pp. 52–3, and see also his other examples on pp. 49–72.

93 Details of the mobile groups are taken from Dick, *Operational Employment*, op. cit.

94 The depth of Montgomery's intended tactical breakthroughs was also very short by Russian standards. It normally stood at about ten kilometres, but was twice that at *Cobra*, and ten times that at *Market-Garden*. In Russia, by contrast, the Germans started to lay out their defences in much greater depth from 1943 onwards, reaching 150km in places towards the end of the war.

Chapter 5

1 E.g. Thompson, W. S., and Frizzell, D. D., *The Lessons of Vietnam* (London, 1977), pp. iv and 277.

2 Ibid., pp. 32–3.

3 This comment was made to the international conference on Vietnam held by the British Commission for Military History at Sandhurst in July 1979. Compare the discussion of regulars using guerilla tactics in Heilbrunn, O., *Warfare in the Enemy's Rear* (London, 1963).

4 Thompson & Frizzell, op. cit., pp. 200–17.

5 Ibid., pp. 89–96.

6 Berger, C., ed., *The USAF in South East Asia* (Office of Airforce History, Washington D.C., 1977).

7 Starry, D. A., *Mounted Combat in Vietnam* (Dept. of the Army *Vietnam Studies* series, Washington D.C., 1978), p. 3.

8 Dickson, P., *The Electronic Battlefield* (London, 1976), p. 82.

9 Ploger, R. R., *US Army Engineers, 1965–70* (*Vietnam Studies*, Washington D.C., 1974), pp. 95–104.

10 Tolson, J. J., *Airmobility 1961–71* (*Vietnam Studies*, Washington D.C., 1973), pp. 12, 41 and 200.

11 Marshall, S. L. A., *Battles in the Monsoon* (New York, 1967), p. 66.

12 Starry, op. cit., p. 160.

Notes

13 Dickson, op. cit., pp. 60–64, and Ott, D. E., *Field Artillery 1954–73* (*Vietnam Studies*, Washington D.C., 1975), p. 181 ff.

14 The name of these tactics was later changed several times in a search for euphemism; Palmer, D. R., *Summons of the Trumpet* (San Rafael, California, 1978), p. 134.

15 Marshall, S. L. A., *Ambush* (New York, 1969), chapter entitled 'The Perfect Deadfall', p. 131.

16 West, F. J., *Small Unit Action in Vietnam* (US Marine Corps, Washington D.C., 1966), p. 95. Compare Albright, J., Cash, J. A., and Sandstrum, A. W., *Seven Firefights in Vietnam* (*Vietnam Studies*, Washington D.C., 1970), Chapter 5, 'Three Companies at Dak To'.

17 *Battles in the Monsoon*, op. cit., pp. 334 and 101; and Marshall, S. L. A., *Bird, the Christmastide Battle* (New York, 1968), p. 152.

18 Ott, op. cit.; and *Battles in the Monsoon*, op. cit., pp. 84 and 91.

19 *Battles in the Monsoon*, op. cit., p. 67.

20 Marshall, S. L. A., *West to Cambodia* (New York, 1968), p. 221.

21 *Bird, the Christmastide Battle*, op. cit., p. 110. Helicopter operations as seen by the pilot are well explained in Mason, R., *Chickenhawk* (first published 1983, Penguin edn., London 1984).

22 West, op. cit., p. 103.

23 Starry, op. cit., p. 83.

24 Ibid., pp. 79–82.

25 Ott, op. cit., p. 54.

26 Katcher, P., *Armies of the Vietnam War, 1962–75* (London, 1980), p. 5.

27 Caputo, P., *A Rumor of War* (London, 1977), p. 81.

28 Tolson, op. cit., p. 252.

29 Palmer, op. cit., p. 97.

30 West, op. cit., p. 111.

31 *Bird, the Christmastide Battle*, op. cit., p. 169.

32 In West, op. cit.

33 *Battles in the Monsoon*, op. cit., p. 338.

34 Ibid., p. 357.

35 Ibid., p. 218.

36 Little, R. W., 'Buddy Relations and Combat Performance', in Janowitz, M., ed., *The New Military* (New edn., New York, 1969), p. 195.

37 Palmer, op. cit., p. 145.

38 Weller, J., *Weapons and Tactics, Hastings to Berlin* (London, 1966), pp. 116–138. Patton's preference for 'marching fire', as opposed to rushes, is explained in English, op. cit., pp. 133–5.

39 Palmer, op. cit., p. 144.

40 *Battles in the Monsoon*, op. cit., p. 95.

41 *Ambush*, op. cit., p. 94.

42 *Bird, the Christmastide Battle*, op. cit., p. 137.

43 Middleton, D., ed., *Air War Vietnam* (USAF, reprinted in London, 1978), p. 102 ff.

44 Palmer, op. cit., p. 145.

45 Caputo, op. cit., p. 70.

46 *Bird, the Christmastide Battle*, op. cit., p. 24.

47 *Battles in the Monsoon*, op. cit., p. 296.

48 Rowan, R., *The Four Days of Mayaguez* (Norton, New York 1975).

49 A favourite slant is apparently provided in Summers H. G. Jr, *On Strategy: the Vietnam war in context* (Novato, California 1982).

50 The point is worth making, however, that the North Vietnamese were not bombed into abandoning any major national interest, only into agreeing that the Americans should leave Indochina.

51   The contrast between the vocabulary of management and the reality of combat is reported to have come home to McNamara himself in the case of the clearing of Ben Suc village, 1967: Schell, J., *The Village of Ben Suc* (Cape, London 1968).

52   The MLRS, or Multiple Launch Rocket System, is a 'smart', long-range version (about 30km) of the Russian *Katyusha*. The Assault Breaker system was originally designed to see deep into the enemy's territory with realtime aerial surveillance, then to guide still bigger rockets on to a target, finally 'closing the combat loop' by scattering a cluster of homing warheads – top-attack anti-tank 'pucks', mines or airfield denial munitions. There is a bewildering range of other ET systems available, including surveillance equipment, target designators, fire control computers and warheads. Some of them use mortars or artillery pieces for delivery, instead of rockets: see general listing of weapons in Barnaby, F., and ter Borg, M., *Emerging Technologies and Military Doctrine* (Macmillan, London 1986), pp. 278–303.

53   TRADOC is the army's Training and Doctrine Command, responsible for analysing weapons and tactics, and producing tactical manuals.

54   See Dickson, op. cit., *passim*, for the electronic battlefield and Vietnam: the pages of *Military Review* for subsequent discussion of FM100-5 – e.g. Wagner, R. E., *Active Defense and All That* in vol. 60, no. 8, August 1980, p. 4; or McCaffrey, B. R., *The Battle on the German Frontier* in vol. 62, no. 3, March 1982, p. 62.

Two strong champions of relatively static firepower-based defence were Hannig, N., *The Defence of Western Europe with Conventional Weapons* in *International Defense Review* vol. 15, no. 3, April 1982, pp. 1439–1442; and Mearsheimer, J. J., *Maneuver, Mobile Defense and the NATO Central Front* in *International Security* vol. 6, no. 3, winter 1981–2, pp. 104–22, and *Why the Soviets Can't Win Quickly in Central Europe* in *International Security* vol. 7, no. 1, Summer 1982, pp. 3–39. Similar 'engineer centred' views are in the unpublished British thesis of Alford, J. R., *Mobile Defense, the Pervasive Myth – an historical investigation* (King's College, London 1977).

55   A very important document issued by TRADOC, on 4 September 1981, was entitled *Airland Battle 2000*. It went beyond FM100-5 to look at the implications of still more high-tech weapons: see Ramon Lopez, *The Airland Battle 2,000 Controversy – Who is being short-sighted?* in *International Defense Review*, October 1983, pp. 1551–6. Similar futurism is in D. R. Cotter's *Potential Future Roles for Conventional and Nuclear Forces in Defense of Western Europe* in *Strengthening Conventional Deterrence in Europe*, the report of the European Security Study (Macmillan, London 1983), pp. 209–53.

56   Starry, D. A., *Extending the Battlefield* in *Military Review*, vol. 61, no. 3, March 1981, pp. 32–50. Compare his stress on mobility in *Mounted Combat in Vietnam*, op. cit.

57   FM 100-5 consists of general operational guidelines for a specifically US army conventional land battle at corps level, in any theatre – i.e. it does not overrule NATO arrangements or strategic directives for Germany. The 1982 version was toned down a little in 1986, but its basic emphasis remained the same: see many articles in *Military Review* throughout the 1980s, particularly Richardson, W. R., *FM100-5: the Air-Land Battle in 1986* in vol. 66, no. 3, March 1986, pp. 4–11, and related discussion in the rest of that issue; also Alcala, R. H., *The United States and the Future of Land Warfare: the AirLand Battle* in Pfaltzgraff, R. L., Jr, Ra'anan, U., Shultz, R. H., and Lukes, I., eds., *Emerging Doctrines and Technologies* (Lexington, Mass. 1988), pp. 173–87.

58   This point is elaborated, and a suitable force structure suggested, in Elmar Dinter and my *Not Over by Christmas* (Bird, Chichester 1983) pp. 14–16 ff.

59   USA alone of the NATO countries favours large-scale helicopter assaults on the enemy's side of the front line. Many notes of caution have been sounded against ET: see e.g. Gouré, D., and McCormick, G., *PGM: No Panacea* in *Survival*, vol. 22, no. 1, Jan–Feb 1980, p. 15; or Dick, C. J., *Soviet Responses to Emerging Technology Weapons and New Defensive Concepts* in Barnaby and Borg, op. cit., pp. 220–38.

60   The attractions of getting away from force structures designed for rapid 'knife

fighting' are mentioned in Pierre, A. J., ed., *The Conventional Defense of Europe* (Council on Foreign Relations, New York, 1986) especially pp. 141–3, 178–9; but see especially extensive debate in Barnaby and Borg, op. cit., pp. 89–124, 215–50. The interface between NATO doctrines and *Airland Battle* is in Bellamy, C., *The Future of Land Warfare* (Croom Helm, London 1987) pp. 124–48.

61   The high-tech light infantry debate is as flourishing today as ever, as is reflected by many articles in *Military Review*. See also English, op. cit., pp. 185–227; and even a direct link with Sir John Moore in Gates, op. cit., pp. 175–80.

62   Schlemmer, B. F., *The Raid* (MacDonald & Jane's, London 1976) pp. 97–9.

63   Starry, *Mounted Combat in Vietnam*, op. cit., pp. 63–4, 73, 84–5. The Sheridan is discussed on pp. 142 ff.

64   Summers, H. G., Jr, *United States Armed Forces in Europe* in Gann, L. H., ed., *The Defense of Western Europe* (Croom Helm, 1987), pp. 299–307.

65   All this is itemised by 'Cincinnatus' in his *Self Destruction* (Norton, New York 1981).

66   Shy, J., and Collier, W., *Revolutionary War*, in Paret (ed.) *Makers of Modern Strategy*, op. cit., pp. 820–1.

67   Gabriel, R. A., *Military Incompetence; why the American military doesn't win* (Hill & Wang, New York 1985), pp. 149–86.

68   Statistics often quoted in Civil War literature, but most fully presented in Moseley, T. V., *The Evolution of American Civil War Infantry Tactics* (unpublished PhD thesis, University of North Carolina, 1967) pp. 195–212.

69   The official record is that 46,498 Americans died in combat in Vietnam, and 10,388 from accidents or otherwise outside the firing line: Lewy, G., *America in Vietnam* (Oxford UP, New York 1978) pp. 450–1.

70   Gabriel, *Military Incompetence*, op. cit., pp. 179–81, suggests this was the number of helicopters lost since lower, officially-quoted figures did not include the helicopters lost by special forces teams before H Hour. Less credibly, Gabriel goes on to suspect that helicopter losses in Vietnam may have been as high as 10,000 rather than the 4,000 normally accepted.

71   Gabriel, op. cit., p. 184.

72   E.g. Creveld's conclusions in *Fighting Power*, op. cit., pp. 164 ff.

73   Montgomery, *El Alamein to the River Sangro*, op. cit., p. 38; d'Este, op. cit., p. 86.

74   E.g. Caputo, op. cit., part I, *The Splendid Little War*; Mason, op. cit., pp. 21–50.

75   See Schlemmer, B. F., *The Raid*, op. cit.

76   This assumption is apparent in many novels and volumes of memoirs, and see e.g. Baker, M., *Nam* (First published 1981, Sphere edn., London 1982).

77   The US casualty lists for each of the 'battles' in Vietnam were tiny by the normal standards of conventional warfare, with around 300 dead at the Ia Drang Valley in 1965, or around 350 at Khe Sanh in 1968. The total US casualties in even the biggest battles were between one and two thousand, and no unit larger than a company was ever overrun. This contrasts with the French loss of some 19,000 at Dien Bien Phu, of which more than a third were killed and wounded.

These considerations did not, of course, apply to either the North or South Vietnamese. The obsessively Americano-centric nature of much of the post-war discussion is far from lost on the many ex-ARVN émigrés now living in the West.

78   The phrase is Joseph Conrad's, climaxing his novella *Heart of Darkness*. It was borrowed by Francis Ford Coppola for his 1979 Vietnam film *Apocalypse Now*; and put on the lips of Marlon Brando.

79   Keegan, op. cit.

80   Extensively discussed in Creveld, Gabriel and Herzog.

81   Manning, F. J., *Continuous Operations in Europe: Feasibility and the Effects on Leadership and Training* in *Parameters*, vol. 9, no. 3, autumn 1979; Hunt-Davis, M. G.,

and Freeman, D. M., *Continuous Operations* in JRUSI, vol. 125, no. 3, September 1980, p. 11.
82 Lopez, op. cit., p. 1554.
83 Dinter, *Hero or Coward*, op. cit.

Chapter 6

1 Graham and Bidwell, op. cit., pp. 233–8, for British attempts to collect data and update tactical doctrine during WWII – an undertaking in which the West has always been outclassed by the Red Army. Pfaltzgraff, et al., op. cit., p. 184, for the US army's Fort Leavenworth 'lessons-learned' cell.
2 Pride in the 1967 'revolutionary' new tactic of leaving the armour unsupported is exhibited in Luttwak, E., and Horrowitz, D., *The Israeli Army* (Lane, London 1975). For an example of how it was still being used – unsuccessfully – even at the very end of the October War, see Rogers, G. F., *The Battle for Suez City* in *Military Review* vol. 59, no. 11, November 1979, pp. 27–33.
3 Ra'anan, U., 'The New Technologies and the Middle East: '"Lessons" of the Yom Kippur War and Anticipated Developments' (in Kempt, G., et al., eds., *The Other Arms Race*, Lexington, Massachusetts, 1975), pp. 79–90. General histories of the war include Palit, *Return to Sinai* (London, 1974), and Herzog, C., *The War of Atonement* (Weidenfeld & Nicolson, London 1975).
4 For technology see *Lessons of Lebanon* in *Defence Attaché* no. 4, 1982, p. 23 ff; and Bellamy, *The Future of Land Warfare* (Croom Helm, London 1987) pp. 27–32, 196–8. Casualties are listed in Herzog, C., *The Arab-Israeli Wars* (Arms & Armour, London 1982) p. 361. Israel lost 480 killed, 2,611 wounded or prisoner, which compare with almost 3,000 dead in 1973, and some 6,000 dead in 1948: ibid., pp. 374, 106.
5 An overview of the war is in Gabriel, R. A. *Operation Peace for Galilee* (Hill & Wang, New York 1984), with a summary of lessons on pp. 191–213.
6 Prosch, G. G., *Israeli Defense of the Golan, an interview with Brigadier General Avigdor Kahalani* in *Military Review* vol. 59, no. 10, October 1979, pp. 2–13.
7 Bellamy, C., *The Future of Land Warfare*, op. cit., p. 291.
8 Chris Bellamy does just this in his otherwise exemplary and most penetrating book (ibid., p. 42). He thereby perhaps betrays that his primary interest lies rather with the technological hardware than with the human software that must use it?
9 Ibid., pp. 19–26, for a summary of a war that has proved exasperatingly hard to follow through the pages of newspapers.
10 Sources are Chandler, D. G., *The Campaigns of Napoleon* (Weidenfeld & Nicolson, London 1967); Terraine, *The Smoke and the Fire*, op. cit., pp. 44–7; and Ascoli, op. cit., pp. 210, 287.
11 Terraine, ibid; Playfair and Molony, op. cit., pp. 78–9; D'Este, op. cit., p. 517.
12 Herzog, op. cit., with additional material from Carver, M., *Conventional Warfare in the Nuclear Age* in Paret, *Makers of Modern Strategy*, op. cit., pp. 779–814, especially pp. 797, 809.
13 Ardant du Picq, op. cit., p. 126.
14 Ibid., p. 44.
15 *Battles in the Monsoon*, op. cit., p. 349. It has sometimes been remarked that S. L. A. Marshall's occasional slighting comments about Ardant du Picq were far from accidental. His debt to the French writer was profound, but he did not wish it to be noticed.
16 E.g. Baron Larrey, quoted in Chandler, op. cit.
17 E.g. General Sherman after the American Civil War, quoted in Ross, S., *From Flintlock to Rifle, Infantry Tactics 1740–1866* (London, 1979), p. 183. Compare Fuller,

J. F. C., *The Conduct of War, 1789–1961* (London, 1961), pp. 105–6, 120 and 130, for a selection of late nineteenth century obituaries on the bayonet.

18   Most of these developments are summarised in Bellamy, *The Future of Land Warfare*, op. cit., pp. 177–273, and Bailey, J. B. A., *Field Artillery and Firepower* (The Military Press, Oxford 1989), pp. 303–36.

19   Bellamy, *The Future of Land Warfare*, op. cit., p. 198. For a suggested way to employ lighter formations in a dispersed battle, see my essay *Countering Surprise by Mobility* in *The Sandhurst Journal* vol. I, no. 1, autumn 1989.

20   Bellamy, *Red God of War*, op. cit., Chapter 3; Bailey, op. cit., *passim*.

21   German strategic geography is well described in Faringdon, Hugh, *Confrontation* (RKP, London 1986), pp. 249–315.

Bibliography

A selection of the more interesting and accessible works consulted.

Histories, Commentaries and Technical Works

General and Strategic Thought

C. von Clausewitz, *On War* (trans J. J. Graham, new edn, 3 vols, London 1940)

J. Colin, *Les Transformations de la Guerre* (Flammarion, Paris 1911)

T. N. Dupuy, *The Evolution of Weapons and Warfare* (First published 1980; Jane's edn, London 1982)

E. M. Earle, ed., *Makers of Modern Strategy* (Princeton University, 1943)

J. A. English, *On Infantry* (first published Praeger, New York 1981; new edn 1984)

H. Jomini, *Précis de l'Art de la Guerre* (new edn, 2 vols, Paris 1955)

J. Luvaas, *The Education of an Army* (Cassell, London 1965)

E. M. Lloyd, *A Review of the History of Infantry* (Longmans, London 1908)

P. Paret, ed., *Makers of Modern Strategy* (Princeton University, 1986)

S. Ross, *From Flintlock to Rifle, Infantry Tactics 1740–1866* (London 1979)

H. Strachan, *European Armies and The Conduct of War* (Allen & Unwin, London 1983)

J. Terraine, *The Smoke and the Fire* (Sidgwick & Jackson, London 1980)

T. Wintringham and J. N. Blashford-Snell, *Weapons and Tactics* (Pelican edn. London 1973)

Napoleonic Wars and Earlier

C. J. J. J. Ardant du Picq, *Battle Studies* (trans J. N. Greely and R. C. Cotton, Macmillan, New York 1921)

206

Bibliography

J. Arnold, *Column and Line in the Napoleonic Wars. A Reappraisal* in *Journal of the Society for Army Historical Research* vol. LX, 1982, pp.196–208

E. Baker, *Remarks on the Rifle* (first published Brighton 1805; 11th edn, London 1835, reprinted Standard publications, Huntington, West Virginia, n.d: c. 1960?)

F. C. Beatson, *With Wellington in the Pyrenees* (Goschen, London n.d., 1914?)

Bressonnet, *Études Tactiques sur la Campagne de 1806* (Chapelot, Paris 1909)

C. B. Brooks, *The Siege of New Orleans* (University of Washington, Seattle 1961)

R. H. Burgoyne, *Historical Records of the 93rd Sutherland Highlanders* (London, Bentley, 1883)

S. Carter 3rd, *Blaze of Glory* (St Martins, New York 1971)

P. Casey, *Louisiana in the War of 1812* (privately printed, Baton Rouge 1963)

D. G. Chandler, *The Campaigns of Napoleon* (Weidenfeld & Nicolson, London 1967)

J. Colin, *La Tactique et la Discipline dans les Armées de la Révolution* (Paris 1902)

C. J. Duffy, *Borodino, Napoleon Against Russia 1812* (Sphere edn, London 1972)

C. J. Duffy, *Austerlitz 1805* (Cooper, London 1977)

C. Duhesme, *Essai Historique sur l'Infanterie Légère* (first published 1814, 3rd edn, Paris 1864)

J. F. C. Fuller, *Sir John Moore's System of Training* (London 1924)

D. Gates, *The British Light Infantry Arm, c. 1790 1815* (Batsford, London 1987)

M. Glover, *Wellington's Peninsular Victories* (Batsford, London 1963)

M. Glover, *Wellington's Army in the Peninsula, 1806–14* (David & Charles, London 1977)

R. Glover, *Peninsular Preparation, The Reform of the British Army, 1795–1809* (London 1963)

P. Griffith, *A Book of Sandhurst Wargames* (Hutchinson, London 1982)

H. de Guibert, 'Essai Générale de Tactique' in Menard, ed., *Guibert, Écrits Militaires 1772–1790* (Paris 1977)

P. Haythornthwaite, *Weapons and Equipment of the Napoleonic Wars* (Blandford, Dorset 1979)

P. Hofschroer, *Prussian Line Infantry, 1792–1815* (Osprey, London 1984)

H. Houssaye, *1814* (73rd edn, Librairie Académique, Paris 1914)

W. James, *Military Occurrences Between Great Britain and America*, vol. II (2 vols, London, 1818)

M. Lauerma, *L'Artillerie de Campagne Française Pendant les Guerres de la Révolution* (Helsinki 1956)

C. A. L. A. Morand, *De l'Armée selon la Charte* (Paris 1829)

G. Nafziger, 'French Infantry Drill, Organization and Training' in *EEL 39*, October 1979, p. 17

A. H. Norris and R. W. Bremner, *The Lines of Torres Vedras* (London 1980)

C. W. Oman, *A History of the Peninsular War* (7 vols, Oxford 1902–30)

C. W. Oman, *Line and Column in the Peninsular War* in *Proceedings of the British Academy*, vol. 4 (London 1910)

P. Paret, *Yorck and the Era of Prussian Reform, 1807–1815* (Princeton University, 1966)

R. Reilly, *The British at the Gates* (Cassell, London 1974)

La Roche Aymon, *Des Troupes Légères* (Paris 1817)

G. von Scharnhorst, *Uber die Wirkung des Feuergewehrs* (Berlin 1813, new edn. Osnabruck 1973)

Mid-Nineteenth Century

M. C. C. Adams, *Our Masters the Rebels* (Harvard UP, Cambridge, Mass. 1978)

D. Ascoli, *A Day of Battle, Mars la Tour, 16th August 1870* (Harrap, London 1987)

E. G. Barrow, *Infantry Fire Tactics* (Hong Kong 1895, reprinted by Wargame Library, Hemel Hempstead, Herts, 1985)

C. C. Buel and R. U. Johnson, eds., *Battles and Leaders of the Civil War* (4 vols, Century, New York 1884 and many later reprints)

T. R. Bugeaud, *Aperçus sur Quelques Détails de la Guerre* (24th edn, Paris 1873)

H. Busk, *The Rifle* (first published in the mid-1850s. 4th edn, London 1859, reprinted Richmond, England 1971)

C. L. Davis, *Arming the Union* (National University, New York 1973)

D. Donald, ed., *Why the North Won the Civil War* (Louisiana University, 1960)

V. J. Esposito, ed., *The West Point Atlas of American Wars* vol. I, *1689–1900* (2 vols, Praeger, New York 1959)

P. Griffith, *French Military Thought and the French Army, 1815–51* (Manchester University, 1989)

P. Griffith, *Rally Once Again* (Crowood, 1987, published in USA by Yale University as *Battle Tactics of the Civil War*, 1989)

E. Hagerman, *The American Civil War and the Origins of Modern Warfare* (Indiana University, 1988)

H. W. Halleck, *Elements of Military Art and Science* (Appleton, New York 1846, reprinted by Greenwood, Westport, Conn. 1971)

E. R. Holmes, *The Road to Sedan, The French Army 1866–70* (Royal Historical Society *Studies in History* Series, no. 41, London 1984)

M. E. Howard, *The Franco-Prussian War* (first published 1960, Fontana edn, London 1967)

F. Lecomte, *Relation Historique et Critique de la Campagne d'Italie en 1859* (Paris 1860)

Captain Loir, *Cavalerie* (Paris, Chapelot 1912)

J. Luvaas, *The Military Legacy of the Civil War – The European Inheritance* (Chicago 1959)

G. McWhiney and P. D. Jamieson, *Attack and Die* (University of Alabama, 1982)

J. K. Mahon, *Civil War Assault Tactics* in *Military Affairs*, vol. 25, summer 1965, pp. 57–68

D. Morris, *The Washing of the Spears* (first published 1965, Sphere edn, London 1968)

T. V. Moseley, *The Evolution of American Civil War Infantry Tactics* (unpublished Ph.D thesis, University of North Carolina, 1967)

C. A. Thoumas, *Les Transformations de l'Armée Française* (2 vols, Paris 1887)

E. Upton, *A New System of Infantry Tactics, Double and Single Rank, adapted to American Topography and Improved Firearms* (Appleton, New York 1867)

J. Weller, 'Shooting Confederate Infantry Arms, Part One' in *The American Rifleman*, April 1954, pp. 43–4

H. Whitehouse, *Battle in Africa* (Fieldbooks, Camberley 1987)

Wilhelm Duke of Württemberg, *The System of Attack of the Prussian Infantry in the Campaign of 1870–71* (trans C. W. Robinson, Aldershot 1871)

H. C. Wylly, *1859, Magenta and Solferino* (Swan Sonnenschein, London 1907)

First World War (and preparations for it)

Anon, *Règlement de Manoeuvre d'Infanterie du 20 Avril 1914* (Chapelot, Paris 1914)

T. Ashworth, *Trench Warfare 1914–1918: The Live and Let Live System* (Macmillan, London 1980)

S. D. Badsey, *Fire and the Sword* (unpublished doctoral thesis, Cambridge 1981, due to be published by Manchester University Press)

W. Balck, *Tactics* (new edn, Posen 1908; trans W. Kreuger, Fort Leavenworth, Kansas 1911; reprinted Greenwood Press, Westport Conn. 1977)

W. Balck, *The Development of Tactics, World War* (trans H. Bell, Fort Leavenworth, Kansas 1922)

S. Bidwell and D. Graham, *Firepower* (Allen & Unwin, London 1982)

G. Blaxland, *Amiens 1918* (Muller, London 1968)

I. S. Bloch, *Modern Weapons and Modern War* (ed. W. T. Stead, London 1899)

G. Blond, *La Marne* (new edn, Livre du Poche, Paris 1962)

British official history, *Military Operations, France and Belgium 1914*, vol. I (Macmillan, London 1923)

Carrias, *La Pensée Militaire Française* (PUF, Paris 1960)

General de Castelli, *Le 8ème. Corps en Lorraine, Aôut – Octobre 1914* (Berger-Levrault, Paris 1925)

Gen H. Colin, *Les Gars du 26ème* (Payot, Paris 1932)

H. Contamine, *La Revanche, 1871–1914* (Berger-Levrault, Paris 1957)

C. Digeon, *La Crise Allemande de la Pensée Française, 1870–1914* (PUF, Paris 1959)

J. Ellis, *The Social History of the Machine Gun* (London 1975)

J. Ellis, *Eye-deep in Hell, The Western Front 1914–18* (London 1976)

French official history, *Les Armées Françaises dans la Grande Guerre*, book I, vol. I (Ministère de la Guerre, Paris 1922)

F. Gambier and M. Suire, *Histoire de la Première Guerre Mondiale*, (2 vols, Fayard, Paris 1968)

R. Girardet, *L'Idée Coloniale en France de 1871 à 1962* (La Table Ronde, Paris 1972)

P-M. de la Gorce, *The French Army* (trans K. Douglas, Weidenfeld & Nicolson, London 1963)

L. de Grandmaison, *Deux conférences faites aux officiers de l'état-major de l'armée: La notion de sûreté, et l'engagement des grandes unités* (Berger-Levrault, Paris 1911)

L. de Grandmaison, *Dressage de l'Infanterie en Vue du Combat offensif* (Berger-Levrault, Paris 1906)

P. Guinard, J-C. Devos and J. Nicot, *Inventaire Sommaire des Archives de la Guerre, Série N 1872–1919* (Renaissance, Troyes 1975)

G. Hanotaux, *L'Enigme de Charleroi* (Edition Française Illustrée, Paris 1917)

J. E. Hicks, *French Military Weapons, 1717–1939* (Flayderman, New Milford, Conn. 1964)

A. Horne, *The Price of Glory, Verdun, 1916* (Penguin edn, London 1964)

General Lanrezac, *Le Plan de Campagne Français et le Premier Mois de la Guerre* (Payot, Paris 1920)

T. T. Lupfer, *The Dynamics of Doctrine: The Changes in German Tactical Doctrine During the First World War* (US Army Command and General Staff College, Fort Leavenworth, Kansas 1981)

L. de Maud'huy, *Infanterie* (Lavauzelle, Paris 1911)

Bibliography

D. Porch, *The March to the Marne, The French Army 1871–1914* (Cambridge University, 1981)
Z. Sternhell, *La Droite Révolutionnaire, 1885–1914* (Seuil, Paris 1978)
E. Swinton, *The Defence of Duffer's Drift* (London 1908)
J. Terraine, *Mons, the Retreat to Victory* (Batsford, London 1960)
T. Travers, 'The Offensive and the Problem of Innovation in British Military Thought, 1870–1915' in *Journal of Contemporary History*, vol. 13, no. 3, July 1978
T. Travers, *The Killing Ground* (Allen & Unwin, London 1987)
General Percin, *Le Combat* (Second edn, Alcan, Paris 1914)
G. C. Wynne, *If Germany Attacks* (London 1940)

Second World War (and preparations for it)

J. A. I. Agar-Hamilton and L. C. F. Turner, *The Sidi Rezeg Battles* (Oxford University Press, Cape Town 1957)
Anon, *Battlefield Tour – Operation Veritable* (BAOR, Germany, December 1947)
Anon, *German Defense Tactics Against Russian Break-Throughs* (US Army Department 20–233, Washington DC, October 1951)
R. G. S. Bidwell, *Gunners at War* (Arms & Armour, London 1970)
C. Bellamy, *Red God of War* (Brassey, London 1986)
A. Brett-James, *Ball of Fire* (Gale and Polden, Aldershot 1951)
'C.A', 'Gazala – The Contrary View' in *Royal Artillery Journal*, March 1988, pp. 40–4
M. Carver, *Tobruk* (first publ. 1964, Pan edn, London 1964)
M. Carver, *El Alamein* (Batsford, London 1962)
M. Carver, *Dilemmas of the Desert War* (Batsford, London 1986)
G. Douhet, *The Command of the Air* (first published 1921, English edn trans D. Ferrari, Coward McCann, New York 1942)
J. F. Dunnigan, ed., *The Russian Front* (Arms & Armour, London 1978)
J. Ellis, *Cassino, the Hollow Victory* (Deutsch, London 1984)
P. Elstob, *Battle of the Reichswald* (Ballantine, London 1970)
C. D'Este, *Decision in Normandy* (Pan edn, London 1984)
A. Farrar-Hockley, *Infantry Tactic, 1939–45* (Almark, London 1976)
N. Frankland, *The Bombing Offensive Against Germany* (Faber, London 1965)
C. de Gaulle, *The Army of the Future* (trans, London 1940)
J. P. Harris, 'The British General Staff and the Coming of War, 1933–9', in *Bulletin of the Institute of Historical Research*, vol. LIX, no. 140, November 1986, pp. 196–211
J. P. Harris, 'British Armour and Rearmament in the 1930s', in *The Journal of Strategic Studies*, vol. 11, no. 2, June 1988, pp. 220–44

J. P. Harris and F. H. Toase, eds, *Modern Armoured Warfare* (forthcoming from Batsford, London 1990)

M. Hastings, *Overlord: D Day and the Battle for Normandy* (Joseph, London 1984)

R. Lewin, *The Life and Death of the Afrika Korps* (Batsford, London 1977)

C. E. Lucas Phillips, *Alamein* (London 1962)

K. Macksey, *Beda Fomm, the Classic Victory* (Ballatine, London 1972)

K. Macksey, *Panzer Division* (Macdonald, London 1968)

K. Macksey, *Tank Force* (Macdonald, London 1970)

K. Macksey, *Tank Warfare, a History of Tanks in Battle* (London 1971)

F. Majdalany, *Cassino, Portrait of a Battle* (Longman, London 1957)

S. L. A. Marshall, *Men Against Fire* (New York 1947)

S. L. A. Marshall, *Battle at Best* (Morrow, New York 1963)

C. Messenger, *The Unknown Alamein* (Ian Allan, London 1982)

A. S. Milward, *The German Economy at War* (Athlone, London 1965)

B. Perrett, *Through Mud and Blood* (Robert Hale, London, 1975)

I. S. O. Playfair, first four volumes of UK official history, *The Mediterranean and Middle East* (the last volume co-authored by C. J. C. Molony, HMSO, London 1954–66)

J. B. Salmond, *The History of the 51st Highland Division 1939–1945* (Blackwood, Edinburgh 1953)

J. L. Scoullar, *Battle for Egypt* (New Zealand official history, Dept. of Internal Affairs, Wellington 1955)

G. L. Verney, *The Desert Rats* (Hutchinson, London 1954)

War Office, *The Armoured Regiment* (Military Training Pamphlet no. 41, London, July 1940)

War Office, *Handling an Armoured Division* (Army Training Instruction no. 3, London, May 1941)

Vietnam

J. Albright, J. A. Cash and A. W. Sandstrum, *Seven Firefights in Vietnam* (*Vietnam Studies* series, Department of the Army, Washington DC 1970)

M. Baker, *Nam* (first published 1981, Sphere edn, London 1982)

'Cincinnatus', *Self Destruction* (Norton, New York 1981)

P. Dickson, *The Electronic Battlefield* (London 1976)

J. J. Ewell and I. A. Hunt, *Sharpening the Combat Edge* (*Vietnam Studies* series, Department of the Army, Washington DC 1974)

W. B. Fulton, *Riverine Operations, 1966–9* (*Vietnam Studies* series, Department of the Army, Washington DC 1973)

J. H. Hay, *Tactical and Material Innovations* (*Vietnam Studies* series, Department of the Army, Washington DC 1974)

G. Lewy, *America in Vietnam* (Oxford University Press, New York 1978)

L. McAulay, *The Battle of Long Tan* (Arrow edn, London 1987)

S. L. A. Marshall, *Battles in the Monsoon* (New York 1967)

S. L. A. Marshall, *Bird, the Christmastide Battle* (New York 1968)

S. L. A. Marshall, *West to Cambodia* (New York 1968)

S. L. A. Marshall, *Ambush* (New York 1969)

S. L. A. Marshall, *The Fields of Bamboo* (New York 1971)

S. L. A. Marshall, *The Raid* (MacDonald & Jane's, London 1976, New York 1968 F. Schlemmer.

D. Middleton, ed., *Air War Vietnam* (USAF, reprinted London 1978)

D. E. Ott, *Field Artillery, 1954–73* (*Vietnam Studies* series, Department of the Army, Washington DC 1975)

D. R. Palmer, *Summons of the Trumpet* (San Rafael, Calif. 1978)

R. R. Ploger, *US Army Engineers, 1965–70* (*Vietnam Studies* series, Department of the Army, Washington DC 1974)

R. Rowan, *The Four Days of Mayaguez* (Norton, New York 1975)

J. Schell, *The Village of Ben Suc* (Cape, London 1968)

D. A. Starry, *Mounted Combat in Vietnam* (*Vietnam Studies* series, Department of the Army, Washington DC 1978)

H. G. Summers Jr, *On Strategy: the Vietnam War in Context* (Novato, Calif. 1982)

W. S. Thompson and D. D. Frizzell, *The Lessons of Vietnam* (London 1977)

J. J. Tolson, *Airmobility, 1961–71* (*Vietnam Studies* series, Department of the Army, Washington DC 1973)

F. J. West, *Small Unit Action in Vietnam* (US Marine Corps, Washington DC 1966)

Other Wars Since 1945, and Into the Future

J. R. Alford, *Mobile Defense, the Pervasive Myth – an historical investigation* (King's College, London 1977)

Anon, 'Lessons of Lebanon' in *Defence Attaché* no. 4, 1982, p. 23ff

Anon, *Operations: FM 100-5, 20 August 1982* (Headquarters, Department of the Army, Washington DC)

J. B. A. Bailey, *Field Artillery and Firepower* (Military Press, Oxford 1989)

F. Barnaby and M. ter Borg, *Emerging Technologies and Military Doctrine* (Macmillan, London 1986)

C. Bellamy, *The Future of Land Warfare* (Croom Helm, London 1987)

E. Dinter and P. Griffith, *Not Over by Christmas* (Bird, Chichester 1983)

ESECS, *Strengthening Conventional Deterrence in Europe, the report of the European Security Study* (Macmillan, London 1983)

H. Faringdon, *Confrontation* (RKP, London 1986)

R. A. Gabriel, *Military Incompetence; why the American military doesn't win* (Hill & Wang, New York 1985)

R. A. Gabriel, *Operation Peace for Galilee* (Hill & Wang, New York 1984)

L. H. Gann, ed., *The Defense of Western Europe* (Croom Helm, 1987)

P. Griffith, 'Countering Surprise by Mobility – a concept for armoured warfare on the Central Front' in *The Sandhurst Journal*, vol. I, no. 1, summer 1989

N. Hannig, 'The Defence of Western Europe with Conventional Weapons' in *International Defense Review* vol. 15, no. 3, April 1982, pp. 1439–42

C. Herzog, *The Arab–Israeli Wars* (Arms & Armour, London 1982)

C. Herzog, *The War of Atonement* (Weidenfeld & Nicolson, London 1975)

W. B. Hopkins, *One Bugle, No Drums: the Marines at Chosin Reservoir* (first publ. 1986, Avon edn, New York 1988)

L. Lavoie, 'Is the Tank Dead?' in *Defense and Diplomacy*, vol. 7, no. 5, May 1989, pp. 16–23.

R. Lopez, 'The Airland Battle 2000 Controversy – Who is being short-sighted?' in *International Defense Review*, 1983 no. 11, pp. 1551–6

E. Luttwak and D. Horrowitz, *The Israeli Army* (Lane, London 1975)

J. J. G. Mackenzie and B. H. Reid (eds), *The British Army and the Operational Level of War* (Tri-Service, London 1989)

S. L. A. Marshall, *Pork Chop Hill* (first publ. 1956, Panther edn, London 1959)

S. L. A. Marshall, *Infantry Operations and Weapons Usage in Korea* (E. C. Ezell ed., Greenhill, London 1988)

J. J. Mearsheimer, 'Maneuver, Mobile Defense and the NATO Central Front' in *International Security* vol. 6, no. 3, winter 1981–2, pp. 104–22

J. J. Mearsheimer, 'Why the Soviets Can't Win Quickly in Central Europe' in *International Security* vol. 7, no. 1, summer 1982, pp. 3–39.

Operations: FM 100-5, 20 August 1982 (Headquarters, Department of the Army, Washington DC)

D. K. Palit, *Return to Sinai* (London 1974)

R. L. Pfaltzgraff Jr, U. Ra'anan, R. H. Shultz and I. Lukes, eds., *Emerging Doctrines and Technologies* (Lexington, Mass. 1988)

A. J. Pierre, ed., *The Conventional Defense of Europe* (Council on Foreign Relations, New York 1986)

G. G. Prosch, 'Israeli Defense of the Golan, an interview with Brigadier General Avigdor Kahalani' in *Military Review* vol. 59, no. 10, Oct. 1979, pp. 2–13

U. Ra'anan, 'The New Technologies and the Middle East: 'Lessons' of the Yom Kippur War and Anticipated Developments' in G. Kemp et al., eds., *The other Arms Race* (Lexington, Mass. 1975), pp. 79–90

W. R. Richardson, 'FM100-5: the Air-Land Battle in 1986' in *Military Review*, vol. 66, no. 3, March 1986, pp. 4–11

G. F. Rogers, 'The Battle for Suez City' in *Military Review*, vol. 59, no. 11, November 1979, pp. 27–33

R. Simpkin, *Race to the Swift* (Brassey, London 1985)

D. A. Starry, 'Extending the Battlefield' in *Military Review*, vol. 61, no. 3, March 1981, pp. 32–50

Combat Stress and Psychology

J. Baynes, *Morale, a Study of Men and Courage* (London 1967)

G. Le Bon, *Psychologie des Foules* (PUF, Paris 1947 edn)

M. Creveld, *Fighting Power* (Arms & Armour, London 1983)

E. Dinter, *Hero or Coward* (Cass, London 1985)

J. Ellis, *The Sharp End of War* (David & Charles, Newton Abbot, Devon 1980)

R. Holmes, *Firing Line* (Cape, London 1985)

J. Keegan, *The Face of Battle* (Cape, London 1976)

Lord Moran, *The Anatomy of Courage* (first published 1985, new edn, Constable, London 1966)

R. A. Nye, *The Origins of Crowd Psychology* (SAGE, London 1975), Chapter 6, pp. 123–153, *Gustave Le Bon and Crowd Theory in French Military Thought Prior to the First World War*

R. Peters, 'The Dangerous Romance; the US army's fascination with the Wehrmacht' in *Military Intelligence*, October–December 1986, pp. 45–8

R. J. Spiller, 'S. L. A. Marshall and the Ratio of Fire', in *RUSI Journal*, vol. 133, no. 4, winter 1988, pp. 63–71

Memoirs and Biographical Materials

Correlli Barnett, *The Desert Generals* (Kimber, London 1960)

J. Brenne, *Un Mobile de l'Armée de Faidherbe* (CRDP, Lille 1972)

P. Caputo, *A Rumour of War* (first publ. 1977, Ballantine edn, New York 1978)

P. Cochrane, *Charlie Company* (Chatto & Windus, London 1977)

R. L. Crimp, *The Diary of a Desert Rat* (Cooper, London 1971)

R. Crisp, *Brazen Chariots* (Muller, London 1959)

F. P. Crozier, *The Men I Killed* (Joseph, London 1937)

H. Curling, ed., *Recollections of Rifleman Harris* (London 1929)

J. D'Arcy Dawson, *Tunisian Battle* (Macdonald, London 1943)

'D. Donovan', *Once a Warrior King* (first publ. 1985, Corgi edn, London 1987)

K. Douglas, *Alamein to Zem Zem* (PL Poetry, London 1946)

J. W. Fortescue, ed., *Memoirs of Sergeant Bourgogne, 1812–13* (London 1926)

G. R. Gleig, *The Subaltern* (first publ. 1825; Dent edn, London n.d. 1949?)

M. Glover, ed., *A Gentleman Volunteer, the letters of George Henell from the Peninsular War, 1812–13* (London 1979)

P. Griffith, ed., *Wellington – Commander* (Bird, Chichester 1985)

H. Guderian, *Panzer Leader* (trans C. Fitzgibbon, new edn, London 1979)

Gurwood, ed., *The Dispatches of the Field Marshall the Duke of Wellington* (London 1835)

I. Hamilton, *A Staff Officer's Scrap Book during the Russo-Japanese War* (2 vols, London 1905)

N. Hamilton *Montgomery*, vol. I, *The Making of a General (3 vols, Hamish Hamilton, London 1981–6)*

W. V. Herbert, *The Defence of Plevna 1877* (London 1895)

M. Herr, *Dispatches* (Picador edn, London 1978)

Horrocks, *Corps Commander* (first publ. 1977, Magnum edn, London 1979)

S. Jary, *18 Platoon* (Sydney Jary, Carshalton Beeches, Surrey 1987)

J. Joffre, *Mémoires du Maréchal Joffre, 1910–17* (2 vols, Plon, Paris 1932)

G. Johnson and C. Dunphie, *Brightly Shone the Dawn, some experiences of the invasion of Normandy* (London 1980)

E. Jünger, *The Storm of Steel* (R. H. Mottram, ed., London 1929)

J. Kincaid, *Adventures in the Rifle Brigade* (J. Fortescue, ed., London 1929)

H. Kippenberger, *Infantry Brigadier* (new edn, London 1961)

B. H. Liddell Hart, ed., *The Letters of Private Wheeler* (Joseph, London 1951)

M. Lindsay, *So Few Got Through* (London 1946)

K. Macksey, *Armoured Crusader* (Hutchinson, London 1967)

E. von Manstein, *Lost Victories* (first publ. 1955, translated Methuen, London 1958)

Martin, *Souvenirs d'un Ex-officier, 1812–15* (Paris 1867)

R. Mason, *Chickenhawk* (first publ. 1983, Penguin edn, London 1984)

J. J. Mearsheimer, *Liddell Hart and the Weight of History* (Cornell University, New York 1988)

F. W. von Mellenthin, *Panzer Battles* (trans Betzler, Cassell, London 1955 and Oklahoma University 1956)

Bibliography

B. Montgomery, *El Alamein to the River Sangro* (first publ. 1948, Grey Arrow edn, 1960)

B. Montgomery, *Normandy to the Baltic* (Hutchinson, London 1946)

G. Orwell, *Homage to Catalonia* (Penguin edn, London 1966)

D. Proctor, *Section Commander* (privately publ., Camberley 1989)

B. H. Reid, *J. F. C. Fuller, Military Thinker* (Macmillan, London 1987)

J. Reid and J. H. Eaton, *The Life of Andrew Jackson* (ed. F. L. Owsley Jr, University of Alabama 1974)

E. Rommel, *Infantry Attacks* (trans G. E. Kiddé, US Marine Corps Association, Quantico Va. 1956)

L. de Saint-Pierre, ed., *Les Cahiers du Général; Brun de Villaret, Pair de France, 1773–1845* (Plon, Paris 1953)

H. W. Schmidt, *With Rommel in the Desert* (Harrap, London 1951)

H. T. Siborne, *Waterloo Letters* (London 1891)

W. Surtees, *Twenty-Five Years in the Rifle Brigade* (first publ. 1833, Military Book Society reprint, London 1973)

R. Trevelyan, *The Fortress, a Diary of Anzio and After* (Penguin edn, London 1979)

A. J. Trythall, *'Boney' Fuller; The Intellectual General* (London 1977)

General Veron, *Souvenirs de ma Vie Militaire – Impressions et Réflexions* (Maugard, Rouen 1969)

J. W. Ward, *Andrew Jackson, Symbol for an Age* (Oxford University Press, New York 1962)

J. Wardrop, *Tanks Across the Desert* (ed. G. Forty, Kimber, London 1981)

M. Warner, *Joan of Arc* (Penguin, London 1983)

G. Wilson, *If You Survive* (Ballantine-Ivy, New York 1987)

T. Wintringham, *English Captain* (London 1939)

W. R. C. Wyne, *Memoir* (privately printed by the family, Southampton c. 1880; to be re-publ. by Fieldbooks, Camberley 1990, ed. by Howard Whitehouse)

Combat Fiction – A Few Personal Perspectives

The Nineteenth Century

It is a pity that Wellington's infantry has been rather badly served in fiction, possibly because it came just a little too early for the great expansion of the novel during the nineteenth century. In modern times there has been no equivalent, for Napoleonic land warfare, to Patrick O'Brien's excellent tales of Nelson's navy. I am reliably assured that the genre picks up nicely a little later in the century with the *Flashman* series

– although I fear that I have not yet made the aquaintance of that particular officer.

To understand the literary basis for the nineteenth century's view of battle, Tolstoy's *War and Peace* should be read in conjunction with Sir Isaiah Berlin's *The Hedgehog and the Fox* (1953), and Georg Lukács' *The Historical Novel* (1937, trans 1962). Tolstoy's *The Raid* offers an early vision of 'the Russian way of war', while Zola's *La Débâcle* gives an all too classic picture of 'the French way' (based on a meticulous reconstruction of Sedan). Stephen Crane's *The Red Badge of Courage* puts inspiration from both Zola and Tolstoy into an American Civil War setting, and Joseph Conrad's *Heart of Darkness* draws inspiration from Crane in turn, with a bleak picture of colonial asset-stripping that must be set beside C. S. Forester's *The Sky and the Forest*.

Captain 'Danrit's' *La Guerre de Demain* is a technically thorough piece of science fiction mixed with revanchist politics, while Viollet le Duc's *Annals of a Fortress* (trans Bucknall, London 1875) is a charmingly fictionalised summary of the history of fortress warfare. But see I. F. F. Clarke's *Voices Prophesying War* (Oxford University, 1966) for a discussion of the whole genre of such future military histories, from Chesney, Verne, Robida and Wells right through to the perennial post-nuclear survival stories of the present day.

The World Wars

Maurice Genevoix' *Ceux de 14* and George Blond's *La Marne* are French equivalents to Solzhenitsyn's *August 1914*. C. S. Forester's *The General* looks at a somewhat later stage of the war, as does the whole corpus of trench horror literature, from E. M. Remarque's *All Quiet on the Western Front* onwards. In *The Great War and Modern Memory* (Oxford UP, 1975) Paul Fussell goes far towards explaining the literary conventions behind these novels and memoirs, although he misses the way in which Remarque's message was twisted into a new mould by such populist story tellers on the Second World War as Sven Hassel and 'Leo Kessler'.

My personal favourites include Ernest Hemingway's military-based novels – *A Farewell to Arms*, *For Whom The Bell Tolls*, and *Over the River and Into the Trees*. See also his well-chosen anthology of war stories, *Men at War* (Fontana edn, London 1966). Fred Majdalany's *Patrol* is a neglected masterpiece describing British infantry fighting during the doldrums of the Second World War, comparable in its way to Evelyn Waugh's more celebrated *Officers and Gentlemen*. Kenneth Macksey's *Battle* is a splendidly detailed and analytical low-level analysis of a 'typical' *bocage* fight in Normandy, 1944; while Len Deighton's *Declarations of War*, no less than his other Second World War fiction, is better than much of the other writing in a similar vein. Unlike any of these, Richard Cox's *Sealion* is based not upon a battle that actually happened, but is surely unique in following the course of a war game about a battle that might have been . . .

Wars Since 1945, and Into the Future

Many Vietnam novels have appeared – particularly in the last few years – accompanied by almost as many Vietnam-related films as there once were Westerns. Unfortunately I stopped following them all as soon as the first edition of the present book had appeared in 1981, so I am badly out of date. Before that time I had found the following books particularly memorable: R. J. Glasser's medical view in *365 Days*; T. O'Brien's *If I Die in a Combat Zone*; and especially James Webb's *Fields of Fire*. Of the films, my favourite is still *The Green Berets*, for the simple reason that it was made at the time and partly in Vietnam itself. It seems to me to possess a subtle documentary quality that has been overlooked, but which its successors lack – although of course the first two *Rambos* are also required viewing for every serious citizen of the modern world.

Vietnam was by no means the only limited war to be fought since 1945. Very many other 'small' or 'cold' wars have attracted literary or cinematographic treatments, including *MASH*, which purported to be about Korea, or *The Virgin Soldiers* about Malaya. Dozens of dime novels, video nasties and arcade games of every sort have more or less explicitly featured clashes between terrorists and straight guys, Cubans and Contras, Libyans and Chadians – or whatever. John Le Carré's work stands as a sensitive monument to the way in which these subjects may properly be treated, but much of the remainder is dross.

When it comes to the future, General Sir John Hackett's *The Third World War* paints a cautionary but ultimately quite optimistic picture of that over-reported future event, while both Shelford Bidwell's NATO view of *World War 3* and Ralph Peters' deep insight into the *Red Army* are cautionary but pessimistic. Kenneth Macksey's *First Clash – Combat Close-up in World War Three* is a detailed low-level view of how the Third World War will appear to Canadian mechanised infantry, and in common with the three aforementioned works is seriously written, by a soldier. It must be admitted, however, that the bestseller lists are otherwise littered with many stories about this war that make military-technical nonsense. I particularly savoured one such offering in which an unfortunate Russian armoured spearhead was told to clear the villages immediately to the West of its border a month before it launched its actual invasion, and then use them to stage dress rehearsals of its surprise attack – but keep it all secret from NATO throughout!

Appendices

I   Glossary of Abbreviations and Technical Terms

ACAV	Armoured Cavalry Assault Vehicle
AHG	*Archives Historiques de la Guerre*
APC	Armoured Personnel Carrier
ARA	Aerial Rocket Artillery
ARVN	Army of the Republic of Vietnam
BEF	British Expeditionary Force
Corps de chasse	Major British armoured formation to complete and exploit infantry breakthroughs in WWII battles
ECM	Electronic Counter-Measures
EEL	Empires, Eagles and Lions magazine
FEBA	Forward Edge of the Battle Area
FOFA	Follow On Forces Attack
FSB	Fire Support Base
HE	High Explosive
JRUSI	Journal of the Royal United Services Institute for Defence Studies
LZ	Landing Zone
MACV	Military Assistance Command, Vietnam
MG	Machine Gun
Mobile Group	Major Soviet armoured formation to complete and exploit infantry breakthroughs in WWII battles.
NCOs	Non Commissioned Officers
NVA	North Vietnamese Army
POW	Prisoner of War
RPV	Remotely Piloted Vehicle
SMG	Sub Machine Gun
TRADOC	Training and Doctrine Command (US Army)
USJ	United Services Journal

II Levels of Action

(*i*) *Strategic*   'National' Strategy embraces a nation's higher military and political objectives. These may well be global in extent, and diplomatic, ideological or economic, rather than military, in nature.

'Theatre' strategy includes the general planning and execution of policy within a particular theatre of war, quite possibly with more than one distinct battle front.

(*ii*) *Operational*   The operational level of war is often referred to in the American military press as 'the lynchpin between tactics and strategy', but its more specific definition is currently a major point of debate. A general indication as to what is meant would perhaps include 'the organisation of military movements for a series of battles on one particular front, usually involving more than one division.'

Operational manoeuvres close to the enemy were once called 'Grand Tactics', but this terminology tends to blur more modern attempts at definition.

(*iii*) *Tactical*   Tactics include everything to do with actually fighting battles, especially within the division.

'Minor tactics' are the principles by which specifically small unit actions are fought, by platoons and companies.

Index

A number of the general subjects discussed in this book are excluded from the index, since they appear in the text so often. These include artillery, bayonets, British troops and tactics, the 'Empty Battlefield', fear, firepower, French troops and tactics, infantry, morale, technology, United States (and Confederate!) troops and tactics, the Napoleonic, Peninsular, First World, Second World and Vietnamese Wars. Conversely, for reasons of space a number of the people and places mentioned only once in passing have been omitted from the index, as have most references to specific military units.

It has also proved impossible to find the initials of a number of nineteenth-century soldiers, since in those days forenames were normally used only within the family circle, and in any case were (in logic quite correctly, as it happens) considered to be far less distinctively personal *outside* the family, than surnames. For example, a regiment might contain seventy 'Paddies' but only one 'Griffith' without an s. However, it is somewhat shocking to the modern historian to find that absolutely no secondary history book – either at the time or subsequently – has managed to reveal the initials of J. A. Margueritte, France's premier cavalier at Sedan . . . or that out of eleven works on 1914 in English consulted by the author, none could supply the initials for (and precious few had apparently even heard of) A. Y. E. Dubail, the spearhead commander at the battle of Sarrebourg, which was itself intended to be the very spearhead and decisive act of at least half of Plan XVII. Alas, it seems that the French von Kluck has found no epitaph.

Note Entries for a particular nation may imply its troops or armies instead of the place itself, and *vice versa*.